ACCELERATING DECARBONIZATION OF THE U.S. ENERGY SYSTEM

Committee on Accelerating Decarbonization in the United States

Board on Energy and Environmental Systems

Division on Engineering and Physical Sciences

Board on Environmental Change and Society

Division of Behavioral and Social Sciences and Education

A Consensus Study Report of

The National Academies of
SCIENCES · ENGINEERING · MEDICINE

THE NATIONAL ACADEMIES PRESS
Washington, DC
www.nap.edu

THE NATIONAL ACADEMIES PRESS • 500 Fifth Street, NW • Washington, DC 20001

This activity was supported by the Alfred P. Sloan Foundation, Heising-Simons Foundation, Quadrivium Foundation, Gates Ventures, ClearPath Foundation, and Incite Labs, with support from the National Academy of Sciences Thomas Lincoln Casey Fund, National Academy of Sciences Arthur L. Day Fund, and National Academy of Sciences Andrew W. Mellon Foundation Fund. Any opinions, findings, conclusions, or recommendations expressed in this publication do not necessarily reflect the views of any organization or agency that provided support for the project.

International Standard Book Number-13:   978-0-309-68292-3

International Standard Book Number-10:   0-309-68292-4

Digital Object Identifier: https://doi.org/10.17226/25932

Additional copies of this publication are available from the National Academies Press, 500 Fifth Street, NW, Keck 360, Washington, DC 20001; (800) 624-6242 or (202) 334-3313; http://www.nap.edu.

Copyright 2021 by the National Academy of Sciences. All rights reserved.

Printed in the United States of America

Suggested citation: National Academies of Sciences, Engineering, and Medicine. 2021. *Accelerating Decarbonization of the U.S. Energy System*. Washington, DC: The National Academies Press. https://doi.org/10.17226/25932.

The National Academies of
SCIENCES · ENGINEERING · MEDICINE

The **National Academy of Sciences** was established in 1863 by an Act of Congress, signed by President Lincoln, as a private, nongovernmental institution to advise the nation on issues related to science and technology. Members are elected by their peers for outstanding contributions to research. Dr. Marcia McNutt is president.

The **National Academy of Engineering** was established in 1964 under the charter of the National Academy of Sciences to bring the practices of engineering to advising the nation. Members are elected by their peers for extraordinary contributions to engineering. Dr. John L. Anderson is president.

The **National Academy of Medicine** (formerly the Institute of Medicine) was established in 1970 under the charter of the National Academy of Sciences to advise the nation on medical and health issues. Members are elected by their peers for distinguished contributions to medicine and health. Dr. Victor J. Dzau is president.

The three Academies work together as the **National Academies of Sciences, Engineering, and Medicine** to provide independent, objective analysis and advice to the nation and conduct other activities to solve complex problems and inform public policy decisions. The National Academies also encourage education and research, recognize outstanding contributions to knowledge, and increase public understanding in matters of science, engineering, and medicine.

Learn more about the National Academies of Sciences, Engineering, and Medicine at **www.nationalacademies.org**.

The National Academies of
SCIENCES · ENGINEERING · MEDICINE

Consensus Study Reports published by the National Academies of Sciences, Engineering, and Medicine document the evidence-based consensus on the study's statement of task by an authoring committee of experts. Reports typically include findings, conclusions, and recommendations based on information gathered by the committee and the committee's deliberations. Each report has been subjected to a rigorous and independent peer-review process and it represents the position of the National Academies on the statement of task.

Proceedings published by the National Academies of Sciences, Engineering, and Medicine chronicle the presentations and discussions at a workshop, symposium, or other event convened by the National Academies. The statements and opinions contained in proceedings are those of the participants and are not endorsed by other participants, the planning committee, or the National Academies.

For information about other products and activities of the National Academies, please visit www.nationalacademies.org/about/whatwedo.

COMMITTEE ON ACCELERATING DECARBONIZATION IN THE UNITED STATES: TECHNOLOGY, POLICY, AND SOCIETAL DIMENSIONS

STEPHEN W. PACALA, NAS,[1] Princeton University, *Chair*
COLIN CUNLIFF, Information Technology and Innovation Foundation
DANIELLE DEANE-RYAN, Libra Foundation
KELLY SIMS GALLAGHER, Tufts University Fletcher School
JULIA HAGGERTY, Montana State University, Bozeman
CHRIS T. HENDRICKSON, NAE,[2] Carnegie Mellon University
JESSE D. JENKINS, Princeton University
ROXANNE JOHNSON, BlueGreen Alliance
TIMOTHY C. LIEUWEN, NAE, Georgia Institute of Technology
VIVIAN LOFTNESS, Carnegie Mellon University
CLARK A. MILLER, Arizona State University
WILLIAM A. PIZER, Duke University
VARUN RAI, University of Texas, Austin
ED RIGHTOR, American Council for an Energy-Efficient Economy
ESTHER TAKEUCHI, NAE, Stony Brook University
SUSAN F. TIERNEY, Analysis Group
JENNIFER WILCOX,[3] Worcester Polytechnic Institute

Staff

K. JOHN HOLMES, Study Director, Board Director/Scholar, Board on Energy and Environmental Systems
ELIZABETH ZEITLER, Associate Director, Board on Energy and Environmental Systems
BRENT HEARD, Program Officer, Board on Energy and Environmental Systems
KASIA KORNECKI, Program Officer, Board on Energy and Environmental Systems
CATHERINE WISE, Associate Program Officer, Board on Energy and Environmental Systems

NOTE: See Appendix B, Disclosure of Conflict(s) of Interest.

---

[1] Member, National Academy of Sciences.
[2] Member, National Academy of Engineering.
[3] Resigned January 2021.

MICHAELA KERXHALLI-KLEINFIELD, Research Associate, Board on Energy and Environmental Systems
REBECCA DEBOER, Research Assistant, Board on Energy and Environmental Systems
HEATHER LOZOWSKI, Financial Business Partner, Board on Energy and Environmental Systems
JENELL WALSH-THOMAS, Program Officer, Board on Environmental Change and Society
CYNDI TRANG, Research Associate, Board on Health Care Services
RANDY ATKINS, Director, Communications/Media, National Academy of Engineering (until July 2020)
MICAH HIMMEL, Senior Program Officer, Transportation Research Board
DAVID BUTLER, Holloman Scholar, National Academy of Engineering

BOARD ON ENERGY AND ENVIRONMENTAL SYSTEMS

JARED COHON, NAE,[1] Carnegie Mellon University, *Chair*
VICKY BAILEY, Anderson Stratton Enterprises
CARLA BAILO, Center for Automotive Research
W. TERRY BOSTON, NAE, GridLiance GP, LLC, and Grid Protection Alliance
DEEPAKRAJ DIVAN, NAE, Georgia Institute of Technology
MARCIUS EXTAVOUR, XPRIZE
KELLY SIMS GALLAGHER, Tufts University Fletcher School
T.J. GLAUTHIER, TJ Glauthier Associates, LLC
NAT GOLDHABER, Claremont Creek Ventures
DENISE GRAY, LG Chem Michigan, Inc.
JOHN KASSAKIAN, NAE, Massachusetts Institute of Technology
BARBARA KATES-GARNICK, Tufts University
DOROTHY ROBYN, Boston University
JOSÉ SANTIESTEBAN, NAE, ExxonMobil Research and Engineering Company
ALEXANDER SLOCUM, NAE, Massachusetts Institute of Technology
JOHN WALL, NAE, Cummins, Inc. (retired)
ROBERT WEISENMILLER, California Energy Commission (former)

Staff

K. JOHN HOLMES, Director/Scholar
ELIZABETH ZEITLER, Associate Director
BRENT HEARD, Program Officer
KASIA KORNECKI, Program Officer
CATHERINE WISE, Associate Program Officer
MICHAELA KERXHALLI-KLEINFIELD, Research Associate
REBECCA DEBOER, Research Assistant
HEATHER LOZOWSKI, Financial Manager
JAMES ZUCCHETTO, Senior Scientist

---

[1] Member, National Academy of Engineering.

BOARD ON ENVIRONMENTAL CHANGE AND SOCIETY

KRISTIE LEE EBI, University of Washington, *Chair*
HALLIE C. EAKIN, Arizona State University
LORI M. HUNTER, University of Colorado, Boulder
KATHARINE JACOBS, University of Arizona
MICHAEL A. MÉNDEZ, University of California, Irvine
RICHARD G. NEWELL, Resources for the Future
ASEEM PRAKASH, University of Washington
MAXINE SAVITZ, NAE,[1] Honeywell, Inc. (former)
MICHAEL P. VANDENBERGH, Vanderbilt University
JALONNE L. WHITE-NEWSOME, Empowering a Green Environment and Economy, LLC
CATHY WHITLOCK, NAS,[2] Montana State University
ROBYN S. WILSON, Ohio State University

Staff

TOBY WARDEN, Director
JENELL M. WALSH-THOMAS, Program Officer
TINA M. LATIMER, Program Coordinator
ADAM JONES, Senior Program Assistant

---

[1] Member, National Academy of Engineering.
[2] Member, National Academy of Sciences.

Preface

Over the past two decades, increased understanding of the severity of impending climate change has coincided with rapid development of non-emitting energy technologies, including significant reductions in their costs. As a result, many nations, states, cities, and companies have recently indicated goals and are developing plans to transition to an energy system that emits zero net anthropogenic greenhouse gases (GHGs), usually by midcentury. This timetable would allow the transition to take advantage of the natural turnover of long-lived capital stock (i.e., the 30-year lifetime of a gas power plant) and is consistent, if adopted globally, with limiting the global temperature increase to substantially less than 2 degrees Celsius.

Because the energy system impacts so many aspects of society, a transition to net zero would have profound implications well beyond climate and energy, including economic competitiveness, increased employment, and improved human health. If done right, a transition to net zero might provide more and better-quality jobs and economic benefits that exceed costs. A transition might also provide an opportunity to eliminate injustices that permeate our current energy system, such as the disproportionate exposure of historically marginalized groups to toxic fossil pollutants. Public support for a decades-long transition could be maintained only by fairly distributing benefits and costs.

Against this backdrop, the National Academies of Sciences, Engineering, and Medicine appointed an ad hoc consensus committee to assess the technological, policy, and social dimensions to accelerate the deep decarbonization of the U.S. economy and recommend research and policy actions in the near to midterm. This interim report focuses on the first 10 years of a 30-year effort—a comprehensive report covering the final two decades will follow in a year. In this interim report, the committee identifies technological actions required during the 2020s to put the United States on a trajectory to net zero by midcentury while still maintaining optionality. Most importantly, the interim report provides a manual for the federal policies needed to enable these technological actions and to build a non-emitting energy system that will strengthen the U.S. economy, promote equity and inclusion, and support communities, businesses, and workers.

The broad scope of this study required a cross-sector analysis and a committee with expertise spanning energy technologies, economics, social sciences, environmental

PREFACE

justice, and policy analysis. The committee worked to produce the interim report from March to October 2020, including innumerable subgroup discussions and three full committee meetings. I would like to thank the committee members for giving so freely of their time, effort, and expertise, especially under the extraordinary circumstances imposed by SARS-CoV-2. Despite a tight timeline and the immensity of the task, the committee members maintained disciplinary rigor while remaining exemplars of interdisciplinary respect. Thanks also to the staff of the National Academies who worked tirelessly to organize us, improve our writing, and help us crystalize our thoughts.

Stephen Pacala, *Chair*
Committee on Accelerating Decarbonization
in the United States: Technological, Policy, and
Societal Dimensions

Acknowledgment of Reviewers

This Consensus Study Report was reviewed in draft form by individuals chosen for their diverse perspectives and technical expertise. The purpose of this independent review is to provide candid and critical comments that will assist the National Academies of Sciences, Engineering, and Medicine in making each published report as sound as possible and to ensure that it meets the institutional standards for quality, objectivity, evidence, and responsiveness to the study charge. The review comments and draft manuscript remain confidential to protect the integrity of the deliberative process.

We thank the following individuals for their review of this report:

Kathleen Araújo, Boise State University
Greg Bertelsen, Climate Leadership Council
Mijin Cha, Occidental College
David L. Greene, University of Tennessee
Noah Kaufman, Columbia University
Kate Konschnik, Duke University
Christopher A. McLean, U.S. Department of Agriculture
Franklin M. Orr, Jr., Stanford University
John Reilly, Massachusetts Institute of Technology
José G. Santiesteban, NAE,[1] ExxonMobil Research and Engineering Company
Emily Schapira, Philadelphia Energy Authority
Kumares C. Sinha, NAE, Purdue University
Addison K. Stark, Bipartisan Policy Center
Nicole Systrom, Sutro Energy Group
Cynthia Winland, Just Transitions Fund

---

[1] Member, National Academy of Engineering.

ACKNOWLEDGMENT OF REVIEWERS

Although the reviewers listed above have provided many constructive comments and suggestions, they were not asked to endorse the conclusions or recommendations of this report nor did they see the final draft before its release. The review of this report was overseen by **Cherry A. Murray,** NAS[2]/NAE, University of Arizona, and **Dan E. Arvizu,** NAE, New Mexico State University. They were responsible for making certain that an independent examination of this report was carried out in accordance with the standards of the National Academies and that all review comments were carefully considered. Responsibility for the final content of this report rests entirely with the authoring committee and the National Academies.

---

[2] Member, National Academy of Sciences.

Contents

Executive Summary	1
Summary	3
1 Motivation to Accelerate Deep Decarbonization	33
Introduction	33
Committee's Approach to the Task Statement	35
Perspectives on the Net-Zero Problem	37
Economics	37
Equity and Fairness	39
Energy Technology	41
Energy Policy	43
Road Map to the Rest of the Report	46
References	51
2 Opportunities for Deep Decarbonization in the United States, 2021–2030	55
Introduction	55
Lessons from Deep Decarbonization Studies and the History of Energy Innovation	59
The First 10 Years: Five Critical Actions	71
Impact on U.S. Energy Expenditures in the 2020s	82
Mobilizing Capital Investment in the 2020s	83
Implications by Sector	92
Conclusion	109
References	110
3 To What End: Societal Goals for Deep Decarbonization	117
Introduction	117
A Social Contract for Decarbonization	120
Leveraging Deep Decarbonization for Economic and Social Innovation	129
Strengthen the U.S. Economy	130
Promote Equity and Inclusion	136
Support Communities, Businesses, and Workers Directly Affected by Transition	143
Maximize Cost-Effectiveness	151
References	154

CONTENTS

4	**How to Achieve Deep Decarbonization**	**163**
	Introduction	163
	Establishing the U.S. Commitment to a Rapid, Just, and Equitable Transition to a Net-Zero Carbon Economy	182
	A Greenhouse Gas Budget for the U.S. Economy	182
	A Price on Carbon with Appropriate Measures to Address Competitiveness and Equity	184
	An Equity and Social Justice Framework	186
	A New Social Contract to Mitigate Harm and Expand Economic Opportunities for Impacted Communities	189
	Setting Rules and Standards to Accelerate the Formation of Markets for Clean Energy That Work for All	191
	A Clean Energy Standard for Electricity	192
	Electrification and Efficiency Standards for Vehicles, Appliances, and Buildings	194
	Improved Regulation and Design of Power Markets for Clean Electricity	197
	Labor Standards for Clean Energy Work	200
	Standards for Corporate Reporting	201
	U.S. Government Procurement Policy and Domestic Clean Energy Markets	203
	Investing in a Net-Zero U.S. Energy Future	206
	Creation of a Green Bank	206
	Invest in New Infrastructure	209
	Invest in Educational Programs for a Clean Energy Workforce	218
	Invest in a Revitalized Manufacturing Sector	221
	Invest in Research, Development, and Demonstration for Technology Innovation and Deployment and Research on Social and Economic Impacts	223
	Invest in Efficiency Improvements for Low-Income Households Through Program Redesign and Expanded Funding	227
	Invest in Electrification of Tribal Lands	228
	Strengthening the U.S. Capacity to Effectively and Equitably Transition to a Clean Energy Future	230
	References	234
APPENDIXES		
A	Committee Biographical Information	245
B	Disclosure of Unavoidable Conflicts of Interest	253

Executive Summary

The world today faces a transformation of its energy system, from one dominated by fossil fuel combustion to one with greatly reduced emissions of carbon dioxide, the primary anthropogenic greenhouse gas (GHG). To help policy makers, businesses, communities, and the public better understand what a transition to net-zero emissions would mean for the United States, the National Academies of Sciences, Engineering, and Medicine convened a committee of experts to investigate how the country could best decarbonize its energy system. The committee was tasked to assess the technological, social, and behavioral dimensions of policies and research activities required over the next 5 to 20 years to put the United States on a path to *net-zero emissions by midcentury*. This interim report of the committee provides a technical blueprint and policy manual for the U.S. energy system over the first critical 10 years of a 30-year effort to transform to net-zero GHG emissions. It focuses on "no-regrets" actions that would be robust to uncertainty about the system's final technological mix, and hedging actions that can keep open as many viable paths to net zero as possible.

Net-zero policy is about more than non-emitting energy technologies, because the manner in which the U.S. economy produces and consumes energy impacts a host of other issues that people care deeply about. The committee recognizes that the energy transition provides an opportunity to build a more competitive U.S. economy, to increase the availability of high-quality jobs, to build an energy system without the social injustices that permeate the current system, and to allow those individuals and businesses that are marginalized today to share equitably in future benefits. To maintain public support through a 30-year transition, the United States will need specific policies to ensure a fair distribution of both costs and benefits. Maintaining public support through a three-decade transition to net zero simply cannot be achieved without the development and maintenance of a strong social contract.

The committee agreed on the following technological and socioeconomic goals for net-zero policy during the 2020s:

Technological Goals

- Invest in energy efficiency and productivity.
- Electrify energy services in transportation, buildings, and industry.
- Produce carbon-free electricity.
- Plan, permit, and build critical infrastructure.
- Expand the innovation toolkit.

Socioeconomic Goals

- Strengthen the U.S. economy.
- Promote equity and inclusion.
- Support communities, businesses, and workers.
- Maximize cost-effectiveness.

This report identifies federal policies to advance these goals and to meet quantitative milestones along the path to net zero. Local, state, and regional policies will be included in the final report. Collectively, the recommended federal policies would catalyze the first 10 years of a transition to net zero and provide the associated environmental, health, and societal benefits, while controlling costs, protecting the competitiveness of the U.S. economy, and compensating for market failures. The policies would also increase the number of high-quality manufacturing jobs, while protecting vulnerable workers and communities, and would reestablish U.S. leadership in energy innovation, manufacturing, and marketing, while building a more just energy system.

Summary

The world has begun a transformation of its energy system from one dominated by fossil fuel combustion to one with net-zero emissions[1] of carbon dioxide (CO_2), the primary anthropogenic greenhouse gas (GHG). This decarbonization is the result of ongoing revolutions in energy technology, public policy, changing economics of energy options, and growing preferences for renewable and zero-carbon supply. In the United States, the energy transformation will require not only a shift from fossil fuel-based to low-carbon sources of energy but also an equally fundamental economic and social transition to strengthen the economy, promote equity and inclusion, and support communities, businesses, and workers.

Examples of the ongoing revolutions in energy technologies are widespread. Low-cost and reliable clean electricity can now become the cornerstone of a net-zero emissions economy, as fuel for electric vehicles, efficient heat pumps, and a source of heat and clean hydrogen for industrial processes. The past decade has seen the levelized cost of wind and solar power drop nearly 70 percent and 90 percent, respectively, while the cost of lithium-ion batteries for electric vehicles dropped by 85 percent. Although the variability of wind and solar makes it impossible to maintain a reliable electricity system with these sources alone, hydropower, energy storage, bioenergy, nuclear energy, geothermal energy, and natural gas with carbon capture and sequestration are available for building a reliable system.

Most near-term emissions reductions during a transition to net zero would come from the electricity sector and the electrification of light- and medium-duty vehicles and home heating. Light-duty transportation and home heating are ready to deliver significant emissions reductions because low-cost, reliable, and clean electricity can be used as fuel for electric vehicles and efficient heat pumps. Substantial improvements in energy efficiency are achievable across all sectors, from buildings to transportation and industry, and can help to meet future demands for energy services cost-effectively. Although technology exists to decarbonize all parts of the energy system, some sectors remain at precommercial or first-of-a-kind demonstration stages and will require significant improvement in cost and performance to become commercially viable. These include aviation, shipping, and industrial subsectors such as steel, cement, and chemicals manufacturing. If innovation fails to provide cost-effective

---

[1] Net-zero emissions are achieved when any CO_2 or other GHG emitted is offset by an equivalent amount of CO_2 removal and sequestration.

alternatives to some of the difficult-to-decarbonize components in time, negative emissions technologies such as direct air capture and storage (DACS), bioenergy with carbon capture and sequestration (BECCS), and enhanced carbon uptake in soils and forests also offer additional options to offset residual emissions from activities that prove more costly to directly decarbonize. The nation has this decade to proactively invest in maturing and improving this suite of solutions and to ensure that as many as possible are prepared for widespread use in the 2030s and 2040s.

This energy transformation is central to mitigating climate change. A transition to net-zero emission in the U.S. economy would directly reduce global CO_2 and other GHG emissions by approximately 10 percent. The country's leadership and innovation capabilities can have an even greater global impact by helping build a suite of affordable clean energy and climate mitigation solutions for export and use around the world. A transition to net zero in the United States would nearly eliminate adverse health impacts of fossil fuel use, which may be responsible for half a million premature deaths or more over the next decade—public health impacts that fall disproportionately on low-income communities and communities of color. Recent polling indicates that a clear majority of Americans now support action to control the country's anthropogenic GHG emissions, as do large majorities of citizens in most other countries.

Given these opportunities, a large and growing number of countries, states, cities, and corporations have pledged to reduce their net GHG emissions to zero over the next 30 years. Although some groups call for a shorter or longer transition period, most target net-zero emissions by 2050 because if this goal is adopted globally, future warming would be limited to a target of 1.5 degrees Celsius. A quicker transition would require expensive replacement of long-lived capital assets before the end of their useful lives. Most proposals call for net-zero emissions with carbon sinks rather than zero emissions because some emissions sources are likely to be too difficult or expensive to mitigate with current and projected technology.

To help policy makers, businesses, communities, and the public better understand what net zero would mean for the United States, the National Academies of Sciences, Engineering, and Medicine convened a committee of experts to investigate how the United States could best decarbonize its energy system. This committee's statement of task (shown in Chapter 1, Box 1.2) called for the committee to "assess the technological, policy, social, and behavioral dimensions to accelerate the decarbonization of the U.S. economy" and "focus its findings and recommendations on near- and midterm (5–20 years) high-value policy improvements and research investments." The statement of task calls for interim and final reports. This interim report focuses on the electricity, transportation, industrial, and buildings sectors, which comprise most of the energy system, and CO_2 emissions, the GHG with the greatest climate impact. In what follows, "energy system" is used as a shorthand for the union of the electricity, transportation, industrial, and buildings sectors. The committee understands that reaching net zero will require addressing all emissions sectors and GHGs. Its final report will include agriculture emissions, expanded treatment of technologies (e.g., hydrogen, low-carbon fuels, and negative emissions technologies), and policy actors (state, local, private sector, and nongovernmental organizations). It will also consider discussion of wider societal trends, such as changes in economics, demographics, housing patterns, and infectious disease incidents that impact the energy system.

This interim report of the committee provides a technical blueprint and policy manual for the first critical 10 years of a 30-year effort to transform the U.S. energy system to net-zero GHG emissions. It focuses on "no-regrets" actions—essential near-term policies that are valuable under any feasible pathway to a net-zero emissions energy system—and the need for some hedging actions during these first 10 years to maintain optionality in the face of substantial uncertainty. For example, renewable sources of electricity will inevitably play a major role given their current low cost, but there are multiple candidates for zero-carbon firm sources of electricity needed because renewable supplies are intermittent. This implies the need for robust research, development, and demonstration (RD&D) across the range of possible candidates, and infrastructure that is specifically planned to be robust to uncertainty in the final mix deployed. It should also be noted that the committee was specifically not tasked to determine whether the nation *should* pursue deep decarbonization, but rather to evaluate options for decarbonization and the highest-priority actions to pursue, *given* that goal.

Net-zero policy is about more than non-emitting energy technologies, because a host of other issues that people care deeply about are also strongly impacted by the ways the U.S. economy produces and consumes energy. The transition represents an opportunity to build a more competitive U.S. economy, increase the availability of high-quality jobs, build an energy system without the social injustices that permeate our current system, and allow those individuals, communities, and businesses that are marginalized today to share equitably in future benefits. Maintaining public support through a three-decade transition to net zero simply cannot be achieved without the development and maintenance of a strong social contract. This is true for all policy proposals described here, including a carbon tax, clean energy standards, and the push to electrify and increase efficiencies in end uses such as vehicle and building energy use. The United States will need specific policies to engage and cultivate public support for the transition, ensure an equitable and just net-zero energy system, and facilitate the recovery of people and communities hurt by the transition.

GOALS AND POLICIES

The committee agreed on the following five technological goals and four socioeconomic goals for net-zero policy during the 2020s:

Technological Goals

- Invest in energy efficiency and productivity.
- Electrify energy services in transportation, buildings, and industry.
- Produce carbon-free electricity.
- Plan, permit, and build critical infrastructure.
- Expand the innovation toolkit.

Socioeconomic Goals

- Strengthen the U.S. economy.
- Promote equity and inclusion.
- Support communities, businesses, and workers.
- Maximize cost-effectiveness.

Each of these goals is discussed below, with some quantitative targets added. Table S.1 at the end of this summary provides the committee's list of highest-priority federal policies for the next 10 years to put the United States on a net-zero path. The table itemizes the policies or groups of policies that together steer the nation's equitable energy transition toward a net-zero economy. Column 1 lists these policies, which are further summarized in the table's notes, in the discussion at the end of this summary, and in Chapter 4. Every policy receives a score for each of the technological goals (shown in column 2, and described in Chapter 2) and socioeconomic goals (shown in column 3, and described in Chapter 3). These scores represent the consensus judgment of the committee's members. Column 4 identifies the branch of the federal government that would be responsible for the policy, and column 5 specifies the required congressional appropriation, if any. The technological and socioeconomic goals are represented by icons (defined below). Icon shade indicates how important each policy is to achieving the goal: darkest shade indicates highest priority—that the policy is indispensable to achieve the objective; medium shade means that the policy is important to achieve the objective; and lightest shade indicates a supporting role. Absence of an icon indicates that the policy would have a small positive role in achieving the objective (and might in some cases have a small negative impact).

The committee's work has been informed by many analyses that examine the implications for costs of various technologies and policies between now and 2050. The committee recognizes the inherent uncertainties that underpin modeling exercises that attempt to capture conditions over the next 3 decades. Thinking back over the past 3 decades, it would have been hard to imagine the energy system implications of the many social and technological changes that have occurred since 1990. Even the best modeling analyses do not capture the expected and unexpected consequences of structural changes over the next 3 decades in electricity demand, in technology change, in fundamental economic trends, in social values and consumer behavior, and in untold other influences.

With an appropriate degree of humility regarding the influence of as yet unknown changes in consumer behavior, technologies, and other transformational changes, the committee has focused this interim report on no-regrets actions that will position the United States on a path toward a net-zero economy. The committee provides numerous instances where it has quantified the types of actions that need to occur in government policies, federal funding, markets, and behaviors to meet the objective of an economy with net-zero emissions by midcentury. These numbers reflect a combination of measures that could meet the net-zero emissions objective with trade-offs across diverse socioeconomic and technical goals. While the committee recognizes that other pathways could accomplish the same objective, it has tried in this interim report to be clear about how those trade-offs have been balanced.

Technological Goals

As reviewed in Chapter 2, recent techno-economic analyses of the net-zero transition in the United States identify five near-term actions in virtually every study that are critical in the 2020s while not locking in a technological mix that might change because of technological advances or breakthroughs. At the same time, a 30-year transition would require that some significant parts of the transition be completed early, either as critical foundations to facilitate other actions, or because expensive pieces of long-lived emitting capital stock reach the end of their useful lives in the 2020s and need to be replaced with a non-emitting alternative (e.g., a gas furnace replaced by an electric heat pump) to avoid lock-in.

The critical near-term actions to accomplish the five technological goals are listed below, next to the icon that identifies the goal in Table S.1.

Invest in energy efficiency and productivity. Over the next 10 years, energy used for space conditioning and plug loads would be reduced in existing buildings by 3 percent per year and total energy use by new buildings reduced by 50 percent. The rate of increase of industrial energy productivity (dollars of economic output per unit of energy consumed) would be increased from a recent pace of 1 percent per year to 3 percent per year. Note that energy efficiency in transportation, buildings, and industry overlaps with electrification, because switching to electric heat pumps and motors also significantly increases the efficiency of heating and transportation relative to fossil-fueled boilers and internal combustion engines. Further, electrification provides opportunities to install broadband and smart grid technologies that enable demand-side management and grid optimization. Also, improvements in efficiency and productivity help to reduce the power loads for equipment, which can reduce the cost of capital and operations lowering hurdles for electrification in these sectors.

⚡ **Electrify energy services in transportation, buildings, and industry.** The most significant actions to accomplish this goal are as follows: reach zero-emissions vehicles as approximately 50 percent of new vehicle sales across all classes by 2030 (light, medium, and heavy); increase the share of electric heat pumps for heating and hot water to 25 percent of residential and 15 percent of commercial buildings, replacing fossil furnaces and boilers; initiate policies for new construction to be all electric in all practical climate zones; and transition low- to moderate-temperature process heat sources to low-carbon electrical power (e.g., by replacing or supplementing conventional units with electric boilers, heat pumps, or noncontact thermal sources such as infrared or microwave) totaling approximately 10 GW of capacity.

Produce carbon-free electricity. During the 2020s, the nation would need to roughly double the share of electricity generated by non-carbon-emitting sources to roughly 75 percent by 2030. Until 2025, this would require an average pace of wind and solar installation that each year matches or exceeds the record historical yearly deployment of these technologies and accelerates to an even faster pace from 2025 to 2030. Emitting coal plants would continue to retire at the current or an accelerated pace. Existing nuclear plants would be preserved wherever it is possible to continue safe operations. Emitting gas-fired generation would decline 10 to 30 percent by 2030 and total capacity would be roughly flat. Some new gas-fired capacity in certain regions could be built during the 2020s to replace aging assets, including coal, because it is more economical than coal regardless of age and can be used to replace aging assets and where coal retirements require replacement capacity for reliability purposes, and where new gas capacity is prepared to retire by 2050 or retrofit to combust hydrogen or be equipped with carbon capture.

Plan, permit, and build critical infrastructure. Build or upgrade electrical transmission facilities to increase overall transmission capacity (as measured in GW-miles) by as much as 60 percent by 2030 to interconnect and harness low-cost wind and solar power across the country. Accelerate the build-out of the nation's electric vehicle (EV) recharging network, including at least 3 million Level 2 chargers and 120,000 DC fast chargers by 2030. This infrastructure should be a mix of private and public ownership and operation, including fleet operators. Plan and initiate a national CO_2 transport and storage network to ensure that CO_2 can be captured at point sources across the country, including in industry, power generation, and low-carbon fuels production (including hydrogen).

Expand the innovation toolkit. The committee proposes a tripling of federal investment in clean energy RD&D to provide new technological options, to reduce costs of existing options, and to better understand how to manage a socially just energy transition. Innovations that would fundamentally enhance the net-zero transition include next-generation energy systems for transportation, buildings, and industry; improved energy storage and firm low-carbon electricity generation options to complement variable renewable electricity; low-cost zero-carbon fuels including hydrogen from the electrolysis of water or biomass gasification; lower-cost carbon capture and use technologies; and lower-cost direct air capture. Progress is needed in particular on net-zero options for aviation, marine transport, and the production of steel, cement, and bulk chemicals. As important will be innovations in how federal policies and programs support RD&D, particularly for technologies in the demonstration and deployment stages.

Please note that some regulatory reform will be necessary to achieve many of the above technological goals. In particular, timely siting and permitting of the new electricity transmission infrastructure is likely to prove difficult or impossible without regulatory reform. Also, the above goals reflect the committee's judgment that a net-zero energy system able to meet the nation's projected business-as-usual demand for energy services will be much easier to achieve than one requiring dramatic reductions in demand for energy services. Thus, the goals do not include greatly reduced mobility or home size.

Socioeconomic goals

A complete transformation of the energy system would affect most aspects of life in this country, with impacts far beyond the installation of new technologies. The U.S. energy system does not currently serve all Americans well. Historically marginalized and low-income populations have energy bills that they struggle to pay and lack the capital to reap benefits from higher-efficiency technologies. They also suffer disproportionate exposure to health and environmental hazards from power generation and climate change with diminished ability to eliminate or mitigate that exposure, have comparatively little say in decision making about siting of energy infrastructure, and receive a disproportionately small share of financial and other benefits from the energy system.

The United States has long been the world's leading technological innovator, but has not effectively used this advantage to sustain domestic manufacturing that could supply domestic and international markets with low- and zero-carbon energy technologies. The decline of the manufacturing sector has cost the economy high-quality jobs, increased income inequality, and contributed to public dissatisfaction.

One cause for optimism is that the country is the best-resourced nation in the world for a transition to net zero. The United States has abundant solar and wind resources both onshore and offshore. Additionally, 40 million acres already are devoted to producing biofuels. The country has plentiful and economically accessible natural gas and enormous geologic and terrestrial reservoirs for CO_2 sequestration.

A transformation to a net-zero economy could combine these natural assets with the nation's culture of innovation to produce an energy system that ameliorates ongoing social injustices in today's energy system and fairly distributes both opportunities and costs. Studies estimate that the transition could increase net employment in the energy system by roughly 1 million to 2 million jobs domestically over the next decade, although the impacts on the location and other characteristics of employment are complex. The innovation and capital expenditures required for a successful transition could revitalize the U.S. manufacturing and commercialization sectors. But the United States will achieve these benefits only if it has the appropriate policies in place. Otherwise, the transition might exacerbate inequity, concentrate opportunity in the hands of a few, accelerate the offshoring of manufacturing, and fail to mitigate job losses in industries and regions that are left behind.

Chapter 3 describes the four critical socioeconomic goals that net-zero policies should be designed to advance. They are as follows:

Strengthen the U.S. economy. The transition to net zero provides an opportunity to revitalize U.S. manufacturing, construction, and commercialization sectors in clean energy and energy efficiency, while providing a net increase in jobs paying higher wages than the national average. The transition would enhance U.S. leadership in clean energy and climate mitigation solutions for which global demand will reach trillions of dollars over coming decades. The net-zero policy portfolio should be designed to strengthen the U.S. economy, with comprehensive policies that enhance the manufacturing sector and promote the innovations needed during the transition.

Promote equity and inclusion. Policies should promote equitable access to the benefits of net-zero energy systems, including reliable and affordable energy, opportunities to benefit from the best available technology, new employment opportunities, and opportunities for financial returns and wealth creation. Net-zero policy should work to eliminate inequities in the current energy system that disadvantage historically marginalized and low-income populations. Net-zero policy must include regular opportunities for, and responses to, community input, as well as ensure fair access to benefits and fair sharing of costs, for the pragmatic reason that public support must be maintained for decades to complete a successful net-zero transition.

Support communities, businesses, and workers. Any fundamental technological and economic transition creates new opportunities as well as job losses in legacy industries and other associated impacts. In particular, the loss of a critical employer could devastate jobs, tax revenues, and other economic impacts in a community or even in whole regions, unless new opportunities can be attracted to replace it with low-carbon competitive employment in a timely manner. Policies should promote fair access to new long-term employment opportunities, provide financial and other support to communities that might otherwise be harmed by the transition, and ensure that jobs created through the transition are high quality, providing at a minimum a safe and secure working environment, family-sustaining wages and comprehensive benefits, regular schedules and hours, and opportunities for skills development.

Maximize cost-effectiveness. This goal begins with an objective to be accomplished—in this case, achieving a net-zero economy by 2050—and finding the least-cost (or most cost-effective) path to accomplish it. Here, the cost of a particular policy is the material consumption that households must give up, including any changes in taxes or government services, to achieve net-zero emissions. A policy's cost-effectiveness measures how this cost compares to the least-cost alternative that achieves the same net-zero outcome and associated benefits. Cost-effectiveness is important because society has multiple objectives, including material well-being. If the country can avoid spending more than necessary in order to achieve net-zero emissions, additional resources are available for other aspirations. However, cost-effectiveness analysis ignores how costs and benefits are distributed within an economy. A U.S. net-zero policy will necessarily need to balance cost-effectiveness with equity and other goals.

System-Wide Policies

Many of the policies listed in Table S.1 would affect the nation's economic and social systems as a whole, given the pervasive (but often invisible) role of carbon in so many elements of Americans' day-to-day experience. The committee's set of recommended policies include some that address these system-wide impacts, facilitate the net-zero transition as a whole, and help advance most of the technological and social-economic goals.

The policy for a U.S. emissions budget covers CO_2 and other GHG emissions and calls for a *target of net zero in 2050* along with regular review of emissions progress and the tracking of specified milestones for technological and social goals. The committee considers a quantitative budget and regular review to be essential for the nation to keep up with the challenging pace required for the net-zero transition, to point out the need to augment policies where progress lags, and to save money where new innovation obviates the need for continuing standards or incentives or costly solutions in markets.

The committee proposes an **economy-wide price on carbon beginning at $40/t CO_2 and rising by 5 percent per year**. The advantages of an economy-wide price on carbon are that it would unlock innovation in every corner of the energy economy, send appropriate signals to myriad public and private decision makers, and encourage a cost-effective route to net zero. However, assuming that the country implements a carbon price before key trade competitors, a mechanism that levels the playing field for domestic firms and avoids emissions leakage will be necessary. Also, because the direct impacts of an economy-wide price on carbon would fall disproportionately on people with the lowest incomes and the fewest choices, it should be augmented by rebates and by funding programs that promote a fair and just transition. The proposed carbon price is deliberately set at a level that would not by itself cause a 30-year transition to net zero because of concerns about equity, fairness, and competitiveness. For example, the committee was not confident that it could design a package of policies that would address competitiveness and mitigate unfair impacts of a carbon price that starts at or climbs rapidly to $100/t$CO_2$.

In addition, the committee calls for the establishment of entities within the federal government to bring equitable access to economic opportunities and wealth creation during the energy transition. These policies are designed to help achieve diversity and fairness goals and to support workers, families, and communities through the transition. The recommendations include the establishment of a 2-year federal **National Transition Task Force** to evaluate the long-term implications of the transition for communities, workers, and families and identify strategies for ensuring a just transition, and a **White House-level Office of Equitable Energy Transitions** to act on the recommendations of the task force, establish just transition targets, and track progress in achieving them by federal programs. The primary policy to help communities achieve new opportunities or mitigate impending damages is the establishment of a new independent **National Transition Corporation**. The committee debated many alternative mechanisms and chose this option because an independent corporation could take the steady long view required to guide the transition initiatives to success.

Private sources of capital are unlikely to be sufficient to finance the low-carbon economic transition, especially during the 2020s when the effort is new. In order to ensure that capital is available for this transition, the committee calls for the establishment of a **Green Bank to mobilize finance**, initially capitalized at $30 billion. Partial financing by a Green Bank would reduce risk for private investors and encourage rapid expansion of private sources capital. To better align the economy with the risks and benefits of transition policies and climate change, the committee includes a policy to require annual Securities and Exchange Commission (SEC) reporting of these risks and benefits by private companies and their inclusion in stress tests by the Federal Reserve and in all cost-benefit analyses by federal agencies.

The committee recommends a **comprehensive education and training initiative** to provide the workforce required for the transition; to improve the competitiveness of the country's building, manufacturing, and energy sectors; and to fuel future innovation. Education and training are also critical to meet societal objectives by providing fair access to new high-quality jobs.

The committee recommends a number of policies to directly enhance and expand the energy innovation toolkit—most notably by the proposed **tripling of the Department of Energy's (DOE's) funding in low- or zero-carbon RD&D over the next 10 years by Congress**, including increasing the agency's funding of large-scale demonstration projects, and the support for social science research on the social and economic aspects of advancing the transition and ensuring that it is just. Chapters 2 and 3 identify specific research needs, while Chapter 4 includes recommendations that propose an allocation of RD&D funding among agencies and among research and demonstration topics.

Past policy measures accelerated RD&D in wind and solar electricity, while financial incentives created niche markets by allowing still-expensive wind and solar to compete with fossil and other sources. The ongoing competition continuously reduced costs of renewables over time by orders of magnitude and returned many times the federal investment to the U.S. economy. The same can be said of the tax credits, other incentives and other federal RD&D support for the development of unconventional natural gas, LED lights, and many other energy innovations. By offering federal support for net-zero RD&D and early market deployment, the policies proposed in Table S.1 will unleash innovation from many sources, from universities and federal labs, to companies competing to capture emerging markets, and to a parallel search by thousands of communities for the best routes to a just and beneficial transition.

Policies Targeting Specific Economic Sectors or Goals

The proposed carbon price would not be large enough during the 2020s to incentivize the deployment of some non-emitting technologies that have relatively high marginal cost and yet must be deployed early, either because long-lived capital stock needs replacement (i.e., a cement plant) or because delay would make the eventual rate of transition infeasible or more expensive. Thus, the committee developed some of its policies in Table S.1 to target specific energy supply and distribution goals.

The committee proposes, for example, a **clean energy standard for electricity** to ensure that the power sector relies increasingly on non-emitting electricity. It also proposes needed policy reforms governing clean electricity markets, amendments to the Federal Power Act to allow timely siting and permitting of new long-distance transmission, and a program to plan, permit, and install the needed new electric transmission capacity. Last, it proposes accelerated installation of smart electricity meters and an expansion of broadband in rural and low-income households. This will allow the electric system to depend upon expanded flexible demand that is enabled by pricing reforms and metering and information-infrastructure upgrades.

Under the committee's recommendations, electrification of the transportation sector and buildings would primarily be accomplished by **manufacturing and performance standards for electric vehicles and building equipment**. For transportation, these would specify fleetwide emissions standards for new vehicle sales that drop to zero in time for the on-road fleet to meet net-zero goals in 2050, appliance standards for the electrification of building heating and cooling, and policies for accelerating the development of electric vehicle charging infrastructure.

To increase the energy efficiency of buildings during the 2020s, the committee calls for weatherization, retrofits, and other support for low-income households, which would also further diversity and fairness goals, as well as emissions caps and efficiency standards for all federal buildings. Note that whole-building energy efficiency can be improved in a multitude of ways, all of which would be simultaneously nudged by the economy-wide price on carbon.

Last, Table S.1 contains the committee's recommendations for policies that directly or indirectly advance a comprehensive clean-energy industrial policy. These include the following policies:

- Output-based allocations and carbon border adjustments that would accompany the carbon price in order to maintain industrial competitiveness;
- A Green Bank to help finance an expansion of clean industry and clean technology manufacturing;
- Corporate climate risk disclosure rules;
- Wholesale power market reforms;
- Education and training policies for the new energy economy;
- Expanded RD&D;
- Electrification of tribal lands;
- A package of loan guarantees and sunsetting subsidies to support installation of non-emitting industrial equipment (e.g., electric boilers) and expand clean-tech manufacturing;

- A process for planning and initiating a national network to transport and safely store CO_2 captured by industrial sources and perhaps by fossil electricity plants with carbon capture; and
- Procurement and other standards for companies that receive federal funds, including labor standards and Buy America/American policies.

COST ESTIMATES AND CAPITAL REQUIREMENTS

This interim report contains three kinds of estimates: the net present value of the aggregate transition costs, the sum of capital required to build all the new hardware and controls in each sector, and the needed congressional appropriations. It also quotes current costs (i.e., levelized cost of energy) of alternative new resources additions. *It is important to note that only the net present value of aggregate transition costs represents a true cost to the United States.* Capital requirements and congressional appropriations can be considered investments in the country's economy that provide long-term returns to private and public sectors. Of course, all of these estimates are highly uncertain. Additionally, any direct costs are balanced against significant public and private benefits of a net-zero transition. These include the substantial avoided health impacts from air pollution within the United States, new economic and employment opportunities, significant downward pressure on global oil prices, and, if other countries also meet similar emissions reductions goals, the avoidance of a substantial portion of planet-altering climate change-related damages to the country that are not already inevitable even with a transition to net zero by midcentury. These could be in the hundreds of billions of dollars annually if estimated health benefits come to fruition and offset some, all, or more than the cost of the transition.

Chapter 2 concludes that the estimated fraction of gross domestic product that the nation would likely spend on energy in a net-zero economy would be smaller than the fraction that the nation has spent on energy in the past, including the past decade (see Chapter 2, Figure 2.3). Studies reviewed in Chapter 2 also estimate total cumulative incremental energy expenditures that average approximately $300 billion through 2030—a roughly 3 percent increase relative to a business-as-usual baseline of approximately $9.4 trillion (net present values of cumulative total expenditures with a 2 percent real social discount rate). It is important to note that these cost estimates do not capture general equilibrium effects, such as changes in global oil prices. Nor do these cost estimates include impacts of changes in the country's balance of trade, which include both positive and negative factors. Last, several of the policies in Table S.1 are designed to reduce or eliminate adverse impacts of costs on trade-impacted firms and low-income households.

TABLE S.1 Summary of Policies Designed to Meet Net-Zero Carbon Emissions Goal and How the Policies Support the Technical and Societal Objectives

Policy	Technological Goals	Socioeconomic Goals	Government Entities	Appropriation, if Any	Notes
Establish U.S. commitment to a rapid, just, equitable transition to a net-zero carbon economy.					
U.S. CO_2 and other GHG emissions budget reaching net zero by 2050.			Executive and Congress	$5 million per year.	Budget is central for imposing emissions discipline, although any consequences for missing the target must be implemented through other policies. Funds are primarily for administration of the budget and data collection and management.
Economy-wide price on carbon.			Congress	None. Revenue of $40/tCO_2$ rising 5% per year, which totals approximately $2 trillion from 2020 to 2030.	Carbon price level not designed to directly achieve net-zero emissions. Additional programs will be necessary to protect the competitiveness of import/export exposed businesses.
Establish 2-year federal National Transition Task Force to assess vulnerability of labor sectors and communities to the transition of the U.S. economy to carbon neutrality.			Congress	$5 million per year.	Task force responsible for design of an ongoing triennial national assessment on transition impacts and opportunities to be conducted by the Office of Equitable Energy Transitions.

Establish White House Office of Equitable Energy Transitions. • Establish criteria to ensure equitable and effective energy transition funding. • Sponsor external research to support development and evaluation of equity indicators and public engagement. • Report annually on energy equity indicators and triennially on transition impacts and opportunities.			Congressional appropriation	$25 million per year, rising to $100 million per year starting in 2025.	Federal office establishes targets and monitors and advances progress of federal programs aimed at a just transition.
Establish an independent National Transition Corporation to ensure coordination and funding in the areas of job losses, critical location infrastructure, and equitable access to economic opportunities and wealth, and to create public energy equity indicators.			Congressional appropriation	$20 billion in funding over 10 years.	Primary means to mediate harms that occur during transition, including support for communities that lose a critical employer, support for displaced workers, abandoned site remediation, and opportunities for communities to invest in a wide range of clean energy projects.

continued

TABLE S.1 Continued

Policy	Technological Goals	Socioeconomic Goals	Government Entities	Appropriation, if Any	Notes
Set rules/standards to accelerate the formation of markets for clean energy that work for all.					
Set clean energy standard for electricity generation, designed to reach 75% zero-emissions electricity by 2030 and decline in emissions intensity to net-zero emissions by 2050.			Congress	None.	
Set national standards for light-, medium-, and heavy-duty zero-emissions vehicles, and extend and strengthen stringency of Corporate Average Fuel Economy (CAFE) standards. Light-duty zero-emission vehicle (ZEV) standard ramps to 50% of sales in 2030; medium- and heavy-duty to 30% of sales in 2030.			Congress	None.	

Set manufacturing standards for zero-emissions appliances, including hot water, cooking, and space heating. Department of Energy (DOE) continues to establish appliance minimum efficiency standards. Standard ramps down to achieve close to 100% all-electric in 2050.			Congress	None.
Enact three near-term actions on new and existing building energy efficiency, two by DOE/Environmental Protection Agency (EPA)[a] and one by the General Services Administration (GSA).			DOE, GSA	None. GSA to set a cap on existing and new federal buildings that declines by 3% per year.
Enact five federal actions to advance clean electricity markets, and to improve their regulation, design, and functioning.[b]			Congress	$8 million per year for Federal Energy Regulatory Commission (FERC) Office of Public Participation and Consumer Advocacy. Two of these actions involve FERC utilizing existing authorities and three involve congressional actions, two directed to FERC and one to DOE.
Deploy advanced electricity meters for the retail market, and support the ability of state regulators to review proposals for time/location-varying retail electricity prices.			Congressional appropriation for DOE	$4 billion over 10 years.

continued

TABLE S.1 Continued

Policy	Technological Goals	Socioeconomic Goals	Government Entities	Appropriation, if Any	Notes
Recipients of federal funds and their contractors must meet labor standards, including Davis-Bacon Act prevailing wage requirements; sign Project Labor Agreements (PLAs) where relevant; and negotiate Community Benefits (or Workforce) Agreements (CBAs) where relevant.		![icons]	Congress	None.	
Report and assess financial and other risks associated with the net-zero transition and climate change by private companies, government agencies, and the Federal Reserve. Private companies receiving federal funds must also report their clean energy research and development (R&D) by category (wind, solar, etc.).	![icons]	![icons]	Congress	None.	Risk disclosures to be included in annual Securities and Exchange Commission (SEC) reports for private companies. Federal Reserve to use climate-related risks in financial stress tests. Federal agencies to include climate-related risks in all benefit cost analyses. All banks to report on comparative financial investments in all energy sources.
Ensure that Buy America and Buy American provisions are applied and enforced for key materials and products in federally funded projects.		![icons]	Congress	None.	

Establish an environmental product declaration library to create the accounting and reporting infrastructure to support the development of a comprehensive Buy Clean policy.			Congressional appropriation for EPA and DOE	$5 million per year.	
Invest (research, technology, people, and infrastructure) in a U.S. net-zero carbon future.					
Establish a federal Green Bank to finance low- or zero-carbon technology, business creation, and infrastructure.			Congressional authorization and appropriation	Capitalized with $30 billion, plus $3 billion per year until 2030.	Additional requirements include public reporting of both energy equity analyses of investment and leadership diversity of firms receiving funds.
Amend the Federal Power Act and Energy Policy Act by making changes to facilitate needed new transmission infrastructure.[c]			Congress	None.	

continued

TABLE S.1 Continued

Policy	Technological Goals	Socioeconomic Goals	Government Entities	Appropriation, if Any	Notes
Plan, fund, permit, and build additional electrical transmission, including long-distance high-voltage, direct current (HVDC). Require fair public participation measures to ensure meaningful community input.[d]			Congressional authorization and appropriation for DOE and FERC	$25 million per year to DOE for planning; $50 million per year for DOE and FERC to facilitate use of existing rights-of-way; finance build through Green Bank; $10 million per year to DOE for distribution system innovations.	Funds provide support for technical assistance to states, communities, and tribes to enable meaningful participation in regional transmission planning and siting activities. Funds to distribution utilities to invest in automation and control technologies.
Expand electric vehicle (EV) charging network for interstate highway system.[e]			Congressional directive to Federal Highway Administration (FHWA) and National Institute of Standards and Technology (NIST); congressional appropriations to DOE	$5 billion over 10 years to expand changing infrastructure.	FHWA to expand its "alternative fuels corridor" program. NIST to develop interoperability standards for level 2 and fast chargers. DOE to fund expansion of interstate charging to support long-distance travel and make investments for EV charging for low-income businesses and residential areas.

Expand broadband for rural and low-income customers to support advanced metering.		Congress to authorize and fund rural electric cooperatives and private companies to offer broadband	$0.5 billion for rural electric cooperatives and $1.5 billion for private companies.	10% of investment costs to expand capabilities of smart grid to underserved areas. Grants or loans to rural electric providers and investment tax incentives to companies, both focused on rural and low-income communities.
Plan and assess the requirements for national CO_2 transport network, characterize geologic storage reservoirs, and establish permitting rules.[f] Require fair public participation measures to ensure meaningful community input.		Congressional authorization and appropriation to multiple agencies	$50 million to Department of Transportation (DOT) with other agencies involved for 5-year planning plus $50 million for block grants for community and stakeholder engagement. $10 billion to $15 billion total during the 2020s to DOE, U.S. Geological Survey (USGS), and Department of the Interior (DOI) to characterize reservoirs. Extend 45Q and increase to $70/tCO_2$–$2 billion per year.	Modeling studies and other analysis indicate that significant amounts of negative emissions will be needed to meet net-zero emissions. The CO_2 pipeline network is needed even with 100% non-fossil electric power to enable carbon capture at cement and other industrial facilities with direct process emissions of greenhouse gases and to enable capture of CO_2 from biomass or via direct air capture for use in production of carbon-neutral liquid and gaseous fuels.

continued

TABLE S.1 Continued

Policy	Technological Goals	Socioeconomic Goals	Government Entities	Appropriation, if Any	Notes
Establish educational and training programs to train the net-zero workforce, with reporting on diversity of participants and job placement success.[g]			Congressional appropriations to Department of Education, DOE, and National Science Foundation (NSF)	$5 billion per year for GI Bill-like program. $100 million per year for new undergraduate programs. $50 million per year for use-inspired and $375 million per year for other doctoral and postdoctoral fellowships. Eliminate visa restrictions for net-zero students. $7 million over 2020–2025 for the Energy Jobs Strategy Council.	Fields covered include science, engineering, policy, and social sciences, for students researching and innovating in low-carbon technologies, sustainable design, and the energy transition.
Revitalize clean energy manufacturing.[h]			Congressional appropriation and direction of Green Bank and U.S. Export-Import Bank	Manufacturing subsidies for low-carbon products starting at $1 billion per year and phased out over 10 years. No additional appropriation required for loans and loan guarantees from Green Bank and Export-Import Bank.	Export-Import Bank should make available at least $500 million per year in low-carbon product and clean-tech export financing and eliminate support for fossil technology exports.

Increase clean energy and net-zero transition research, development, and demonstration (RD&D) that integrates equity indicators.[i]		Congressional appropriation for and directions to DOE and NSF	DOE clean energy RD&D triples from $6.8 billion per year to $20 billion per year over 10 years. DOE funds studies of policy evaluation at $25 million per year and regional innovation hubs at $10 million per year; DOE- and NSF-funded studies of social dimensions of the transition should be supported by an appropriation of $25 million per year.	Establish criteria for receiving funds on equity analysis, appropriate community input, and leadership diversity of companies applying for public investments. DOE to report on equity impacts and diversity of entities receiving public funds.
Increase funds for low-income households for energy expenses, home electrification, and weatherization.		Congressional appropriation	Increase Weatherization Assistance Program (WAP) funding to $1.2 billion per year from $305 million per year. Direct the Department of Health and Human Services (HHS) to increase state's share of Low Income Home Energy Assistance Program (LIHEAP) funds for home electrification and efficiency.	

continued

TABLE S.1 Continued

Policy	Technological Goals	Socioeconomic Goals	Government Entities	Appropriation, if Any	Notes
Increase electrification of tribal lands			Congressional appropriation to DOE and U.S. Department of Agriculture (USDA)	$20 million per year for assessment and planning through DOE Office of Indian Energy Policy (DOE-IE) and USDA Rural Utilities Service (USDA-RUS); expand DOE-IE to $200 million per year.	Increase direct financial assistance for the build-out of electricity infrastructure through DOE-IE grant programs.
Assist families, businesses, communities, cities, and states in an equitable transition, ensuring that the disadvantaged and at-risk do not suffer disproportionate burdens.					
Please note that the primary policies targeting fairness, diversity, and inclusion during the transition are establishing the Office of Equitable Energy Transitions and the National Transition Corporation, which are the fourth and fifth policies in this table.					
Establish National Laboratory support to subnational entities for planning and implementation of net-zero transition.			Congressional appropriation	Additional funding to national laboratories' annual funding commencing at the level of $200 million per year, rising to $500 million per year by 2025, and $1 billion per year by 2030.	To establish a coordinated, multi-laboratory capability to provide energy modeling, data, and analytic and technical support to cities, states, and regions to complete a just, equitable, effective, and rapid transition to net zero.

Establish 10 regional centers to manage socioeconomic dimensions of the net-zero transition.[j]		Congressional authorization and appropriations to DOE	$5 million per year for each center; $25 million per year for external research budget to provide data, models, and decision support to the region.	Coordinated by the Office of Equitable Energy Transitions.
Establish net-zero transition office in each state capital.		Congressional appropriations	$1 million per year in matching funds for each state.	Coordinate state's effort with federal and regional efforts.
Establish local community block grants for planning and to help identify especially at-risk communities. Greatly improve environmental justice (EJ) mapping and screening tool and reporting to guide investments.		Congressional appropriations to DOE	$1 billion per year in grants administered by regional centers.	Required to qualify for funding from the National Transition Corporation. Block grant funding requires inclusive participation and engagement by historically marginalized and low-income groups.

continued

TABLE S.1 Continued

KEY TO ICONS
DARK GREEN icon indicates that the policy is highest priority and indispensable to achieve the objective. **MEDIUM GREEN** icon indicates that the policy is important to achieve the objective. **LIGHT GREEN** icon indicates that the policy would play a supporting role. No icon indicates that the policy would have at most a small positive role in achieving the objective (and might in, some cases, have a small negative impact on the objective).
Technological Goals
Invest in energy efficiency and productivity. Examples include accelerating the rate of increase of industrial energy productivity (dollars of economic output per energy consumed) from the historic 1% per year to 3% per year.
Electrify energy services in transportation, buildings, and industry. Examples include, by 2030, moving half of vehicle sales (all classes combined) to EVs, and deploying heat pumps in one-quarter of residences.
Produce carbon-free electricity. Roughly double the share of electricity generated by carbon-free sources from 37% to 75%.
Plan, permit, and build critical infrastructure. Build critical infrastructure needed for the transition to net zero, including new transmission lines, an EV charging station network, and a CO_2 pipeline network.
Expand the innovation toolkit. Triple federal support for net-zero RD&D.
Socioeconomic Goals
Strengthen the U.S. economy. Use the energy transition to accelerate U.S. innovation, reestablish U.S. manufacturing, increase the nation's global economic competitiveness, and increase the availability of high-quality jobs.
Promote equity and inclusion. Ensure equitable distribution of benefits, risks, and costs of the transition to net zero. Integrate historically marginalized groups into decision making by ensuring adherence to best-practice public participation laws. Require that entities receiving public funds report on leadership diversity to ensure nondiscrimination.
Support communities, businesses, and workers. Ensure support for those directly and adversely affected by the transition.
Maximize the cost-effectiveness of the transition to net zero.

[a] Direct DOE/EPA to expand outreach of and support for adoption of benchmarking and transparency standards by state and local government through the expansion of Portfolio Manager. Direct DOE/EPA to further investigate the development of model carbon-neutral standards for new and existing buildings that, in turn, could be adopted by states and local authorities. Policies targeting retrofits of existing buildings will be in the final report.

[b] FERC should work with regional transmission organizations (RTOs) and independent system operators (ISOs) to ensure that markets in all parts of the country are designed to accommodate the shift to 100% clean electricity on the relevant timetable. Congress should clarify that the Federal Power Act does not limit the ability of states to use policies (e.g. long-term contracting with zero-carbon resources procured through market-based mechanisms) to support entry of zero-carbon resources into electric utility portfolios and wholesale power markets. Congress should further direct FERC to exercise its rate-making authority over wholesale prices in ways that accommodate state action to shape the timing and character of the transitions in their electric resource mixes. Congress should reauthorize the FERC Office of Public Participation and Consumer Advocacy to provide grants and other assistance to support greater public participation in FERC proceedings. FERC should direct the North American Electric Reliability Corporation (NERC) to establish and implement standards to ensure that grid operators have sufficient flexible resources to maintain operational reliability of electric systems. Congress should direct and fund DOE to provide federal grants to support the deployment of advanced meters for retail electricity customers as well as the capabilities of state regulatory agencies and energy offices to review proposals for time/location-varying retail electricity prices, while also ensuring that low-income consumers have access to affordable basic electricity service.

[c] (1) Establish National Transmission Policy to rely on the high-voltage transmission system to support the nation's (and states') goals to achieve net-zero carbon emissions in the power sector. (2) Authorize and direct FERC to require transmission companies and regional transmission organizations to analyze and plan for economically attractive opportunities to build out the interstate electric system to connect regions that are rich in renewable resources with high-demand regions; this is in addition to the traditional planning goals of reliability and economic efficiency in the electric system. (3) Amend the Energy Policy Act of 2005 to assign to FERC the responsibility to designate any new National Interest Electric Transmission Corridors and to clarify that it is in the national interest for the United States to achieve net-zero climate goals as part of any such designations. (4) Authorize FERC to issue certificates of public need and convenience for interstate transmission lines (along the lines now in place for certification of gas pipelines), with clear direction to FERC that it should consider the location of renewable and other resources to support climate-mitigation objectives, as well as community impacts and state policies as part of the need determination (i.e., in addition to cost and reliability issues) and that FERC should broadly allocate the costs of transmission enhancements designed to expand regional energy systems in support of decarbonizing the electric system.

[d] (1) Congress should authorize and appropriate funding for DOE to provide support for technical assistance and planning grants to states, communities, and tribal nations to enable meaningful participation in regional transmission planning and siting activities. (2) Congress should authorize and appropriate funding for DOE and FERC to encourage and facilitate use of existing rights-of-way (e.g., railroad; roads and highways; electric transmission corridors) for expansion of electric transmission systems. (3) Congress should authorize and appropriate funding for DOE to analyze, plan for, and develop workable business model/regulatory structures, and provide financial incentives (through the Green Bank) for

continued

TABLE S.1 Continued

development of transmission systems to support development of offshore wind and for development, permitting, and construction of high-voltage transmission lines, including high-voltage direct-current lines.

[e] (1) Congress should direct the Federal Highway Administration (a) to continue to expand its "alternative fuels corridor" program, which supports planning for EV charging infrastructure on the nation's interstate highways, and (b) to update its assessment of the ability and plans of the private sector to build out the EV charging infrastructure consistent with the pace of EV deployment needed for vehicle electrification anticipated for deep decarbonization, the need for vehicles on interstate highways and in public locations or high-density workplaces, and to identify gaps in funding and financial incentives as needed. In coordination with FHWA, DOE should provide funding for additional EV infrastructure that would cover gaps in interstate charging to support long-distance travel and make investments for EV charging for low-income businesses and residential areas. (2) NIST should develop communications and technology interoperability standards for all EV level 2 and fast charging infrastructure.

[f] Extend 45Q tax credit for carbon capture, use, and sequestration for projects that begin substantial construction prior to 2030 and make tax credit fully refundable for projects that commence construction prior to December 31, 2022. Set the 45Q subsidy rate for use equal to \$35/tCO_2 less whatever explicit carbon price is established and the subsidy rate for permanent sequestration to be equal to \$70/tCO_2 less whatever explicit carbon price is established. A hydrogen pipeline network will ultimately also be needed, but, as indicated in Chapter 2, the time pressure to build a national hydrogen pipeline network is less severe than for CO_2. This is because hydrogen production facilities can be located close to industrial hydrogen consumers, unlike CO_2 pipelines, which must terminate in geologic storage reservoirs. Also, hydrogen can be blended into natural gas and transported in existing gas pipelines, and gas pipelines could ultimately be converted to 100% hydrogen.

[g] (1) Congress should establish a 10-year GI Bill-type program for anyone who wants a vocational, undergraduate, or master's degree related to clean energy, energy efficiency, building electrification, sustainable design, or low-carbon technology. Such a program would ensure that the U.S. workforce transitions along the physical infrastructure of our energy, transportation, and economic systems. (2) Congress should support the creation of innovative new degree programs in community colleges and colleges and universities focused uniquely on the knowledge and skills necessary for a low-carbon economic and energy transformation. (3) Congress should provide funds to create interdisciplinary doctoral and postdoctoral training programs, similar to those funded by the National Institutes of Health (NIH), which place an emphasis on training students to pursue interdisciplinary, use-inspired research in collaboration with external stakeholders that can guide research and put it to use in improving practical actions to support decarbonization and energy justice. (4) Congress should provide support for doctoral and postdoctoral fellowships in science and engineering, policy, and social sciences for students researching and innovating in low-carbon technologies, sustainable design, and energy transitions, with at least 25 fellowships per state to ensure regional equity and build skills and knowledge throughout the United States. (5) The Department of Homeland Security (DHS) should eliminate or ease visa restrictions for international students who want to study climate change and clean energy at the undergraduate and graduate levels, where appropriate. (6) Congress should pass the Promoting American Energy Jobs Act of 2019 to reestablish the Energy Jobs Strategy Council under DOE, require energy and employment data collection and analysis, and provide a public report on energy and employment in the United States.

[h] (1) Congress should establish predictable and broad-based market-formation policies that create demand for low-carbon goods and services, improve access to finance, create performance-based manufacturing incentives, and promote exports. Specifically, Congress should provide manufacturing incentive through loans, loan guarantees, tax credits, grants, and other policy tools to firms that are matched with corresponding performance requirements. Subsidies provided directly to manufacturers must be tied to the meeting of performance metrics, such as production of products with lower embodied carbon or adoption of low-carbon technologies and approaches. Specific items could include expanding the scope of the energy audits in the DOE Better Plants program and expanded technical assistance to focus on energy use and GHG emissions reductions at the 1,500 largest carbon-emitting manufacturing plants; supporting the hiring of industrial plant energy managers by having DOE provide manufacturers with matching funds for 3 years to hire new plant energy managers; enabling the development of agile and resilient domestic supply chains through DOE research, technical assistance, and grants to assist manufacturing facilities in addressing supply chain disruptions resulting from COVID-19 and future crises. (2) Congress should provide loans and loan guarantees to manufacturers to produce low-carbon products, ideally through a Green Bank (see Chapter 4). (3) Congress should require the U.S. Export-Import Bank to phase out support for fossil fuels and make support for clean energy technologies a top priority with a minimum of $500 million per year. (4) Congress should create a new Assistant Secretary for Carbon Smart Manufacturing and Industry within DOE.

[i] (1) Congress should triple the DOE's investments in low- or zero-carbon RD&D over the next 10 years, in part by eliminating investments in fossil-fuel RD&D. These investments should include renewables, efficiency, storage, transmission and distribution (T&D), carbon capture, utilization, and storage (CCUS), advanced nuclear, and negative emissions technologies and increase the agency's funding of large-scale demonstration projects. By eliminating investments in non-carbon capture and sequestration (non-CCS) fossil-fuel RD&D, the net increase to the energy RD&D budget will be partially offset. (2) Congress should direct DOE to fund energy innovation policy evaluation studies to determine the extent to which policies implemented (both RD&D investment and market-formation policies) are working. (3) Congress should direct DOE and NSF to create a joint program to fund studies of the social, economic, ethical, and organizational drivers, dynamics, and outcomes of the transition to a carbon-neutral economy, as well as studies of effective public engagement strategies for strengthening the U.S. social contract for decarbonization. (4) Congress should direct DOE to establish regional innovation hubs where they do not exist or are critically needed using funds appropriated under item 1 above. (5) Congress should direct DOE to enhance public-private partnerships for low-carbon energy.

[j] (1) Congress should coordinate federal agency actions at the regional scale through the deployment of federal agency staff to regional offices. (2) Congress should host a coordinating council of regional governors and mayors that meets annually to establish high-level policy goals for the transition. (3) Congress should establish mechanisms for ensuring the effective participation of low-income communities, communities of color, and other disadvantaged communities in regional dialogue and decision making about the transition to a carbon-neutral economy. (4) Congress should provide information annually to the White House Office of Equitable Energy Transitions detailing regional progress toward decarbonization goals and benchmarks for equity.

Chapter 2 also reports that roughly $2 trillion in incremental capital investments must be mobilized over the next decade for projects that come online in 2030 to put the United States on track to net zero by 2050 (average from studies identified in Figure 2.5). These capital investments are not a direct cost borne by either taxpayers or energy consumers. The sum of capital investments that must be mobilized in the 2020s is much larger than the increase in total consumer energy expenditures described above because capital investments are paid back through energy expenditures over many years and because investments in renewable electricity, efficient buildings and vehicles, and other capital-intensive measures offset significant annual expenditures on consumption of fuels. Capital investment estimates are included in the report because policies will be needed to directly finance some projects and de-risk others, given that private capital markets are not currently set up for the net-zero transition.

The committee estimates that $350 billion over a 10-year period in total federal appropriations would be needed to fund the package of net-zero transition policies described above. The carbon price proposed in Chapter 4 would also raise approximately $2 trillion over the decade (2021–2030), providing revenue to fully offset proposed appropriations and provide substantial funds for targeted rebates and other programs to address equity and distributional concerns.

CONCLUSION

A transition to a net-zero economy in the United States by midcentury is technologically feasible, with energy system costs as a share of U.S. gross domestic product that have been manageable over the past decade, but it is on the edge of feasibility. These conditions warrant rapid rates of change and unprecedented levels of funding for RD&D, infrastructure planning, permitting and construction activity, and other changes in public policy and social systems that have to begin immediately across the energy economy, as well as unprecedented actions to build and maintain public support for the net-zero transition.

With an appropriate portfolio of policies, however, the transition will advance a number of national objectives simultaneously: building a more fair and just energy system that works for all Americans, improving the international competitiveness of the economy, revitalizing American manufacturing, and reestablishing leadership in energy innovation and technology. The transition will also provide new high-quality jobs, virtually eliminate the substantial health impacts of fossil fuels, reduce U.S. GHG emissions to zero, enhance the nation's leadership in climate and energy policy, and help catalyze the global transition necessary to avert the most damaging impacts of business-as-usual climate change.

CHAPTER ONE

Motivation to Accelerate Deep Decarbonization

INTRODUCTION

Humanity has already embarked on a transformation of the global energy system that could, upon completion, approach the scale of a second Industrial Revolution. Every year, damages from climate change become better documented and understood, as well as more widespread and severe (IPCC, 2018). Every year, public support for action becomes stronger, both globally and within the United States, as people experience the effects of climate change firsthand (Pew Research Center, 2020). Every year, millions die worldwide, including up to 200,000 Americans, because of pollution caused by producing and combusting fossil fuel (Lelieveld et al., 2019). Every year, non-emitting energy technologies become cheaper and more available (see Chapter 2 of this report). This is why so many nations, states, cities, and companies have committed to replacing our current energy system by midcentury with a system that would emit zero net anthropogenic greenhouse gases (GHGs) (CDP, 2019; U.S. Climate Alliance, 2020; We Are Still In, 2020). Tens of trillions of dollars in costs and revenues hang in the balance, as do living conditions both at home and around the globe.

Many proposals to achieve net zero in the United States have been released, primarily by advocacy groups, political campaigns, and members of Congress. These plans target net-zero rather than zero emissions because some GHG sources would be too disruptive or expensive to eliminate (i.e., some agricultural methane and N_2O; see Box 1.1).[1] Net-zero emissions are achieved when any CO_2 or other GHG emitted is offset by an equivalent amount of CO_2 removal and sequestration. Most plans would offset between 10 and 20 percent of current emissions by negative CO_2 emissions (carbon sinks or carbon removal) of the same magnitude.

---

[1] The focus of the interim report is on reducing CO_2 emissions from the energy system in the United States while recognizing that there are other GHGs that contribute to climate change and that need to be reduced. The use of carbon dioxide equivalent (CO_2e) is a metric for describing the global warming potential of different GHGs in a common unit by defining the number of units of CO_2 that would have the equivalent global warming impact of one unit of another GHG. While simple to describe, GHGs have different atmospheric lifetimes. Osko et al. (2017) discuss the temporal trade-offs inherent in using a single time frame for estimating CO_2e and recommend reporting this metric for multiple time frames.

> **BOX 1.1**
> **CURRENT GREENHOUSE GAS EMISSIONS**
>
> Global anthropogenic emissions of all greenhouse gases (GHGs) amounted to 55 Gt CO_2e/y in 2019, the majority as CO_2 (37 Gt CO_2/y) and the rest as methane, N_2O, and fluorinated gases. Corresponding emissions for the United States were 6 Gt CO_2e/y of all GHGs and 5 Gt CO_2/y. Ninety percent of global CO_2 emissions is caused by fossil fuel combustion (Friedlingstein et al., 2019). The majority of methane and N_2O emissions are agricultural, but approximately one-third of methane emissions represent natural gas that escapes from oil, gas, and coal operations, or that escapes in transportation or storage before being combusted by an end-user (Saunois et al., 2020). Fluorinated gases primarily escape during industrial use and the production and aging of refrigeration and cooling systems. The United States also possesses a large CO_2 sink from its managed forests of approximately 0.7 Gt CO_2/y, which approximately offsets the nation's agricultural emissions (EPA, 2020). Thus, reducing U.S. net emissions to zero over 30 years means that net emissions must be reduced by an average of approximately 0.2 Gt CO_2e/y.

The 30-year time frame of most net-zero proposals comes from two sources. First, global anthropogenic emissions must reach net zero by approximately midcentury to limit climate change to substantially less than 2 degrees Celsius (IPCC, 2018). Second, many energy system and industrial assets discussed in Chapter 2 last for years or even decades, from personal vehicles and natural gas plants to cement facilities and industrial boilers. A transition to net zero is far cheaper if long-lived components are allowed to reach the end of their useful lives before being replaced by non-emitting alternatives, and studies have found that a 30-year horizon for a net-zero transition leverages the normal pace of asset replacement and avoids significant premature retirement of existing assets.

The National Academies of Sciences, Engineering, and Medicine were established to provide expert advice to the nation. This advice is carefully peer reviewed, financially disinterested, apolitical, and nonideological. National Academies committees are chosen to avoid financial, ideological, or political conflicts of interest. To help federal, state, and local policy makers, businesses, and other community leaders and the general public better understand what net zero would mean for the country, the National Academies convened a committee to investigate how the United States could best decarbonize its energy system.

This document offers a technical blueprint and policy manual for the first 10 years of a 30-year effort to replace the current U.S. energy system with one that has net-zero anthropogenic emissions. It begins (in Chapter 2) with an analysis of essential actions that would have to be taken over the next 10 years to make the 30-year objective feasible, while preserving optionality about the mix of technologies in the 2050 energy system to allow for innovation, changes in points of view, and surprises. The committee refers to essential near-term policies that are valuable under any feasible net-zero energy system pathway as "no-regrets" policies. The report then turns (in Chapter 3) to a discussion of societal impacts of our current energy system and the transition to a net-zero system, including how inequities built into our current system could be eliminated, how communities and groups that would otherwise be damaged by the transition could be sustained, how U.S. international economic, political, and technological leadership could be enhanced, and how our domestic manufacturing sector and the high-quality jobs within it could be revitalized. The bulk of the report (Chapter 4) then describes and explains the highest-priority policies for the first 10 years of a 30-year transition. A more comprehensive report covering the full 30 years will follow in a year.

COMMITTEE'S APPROACH TO THE TASK STATEMENT

National Academies committees are bound by their statements of task; the statement of task for this committee is shown in Box 1.2. The committee interpreted "deep decarbonization" in the statement of task to mean net zero by 2050, because of this target's widespread use. However, because this interim report focuses on actions that would be needed in the next 10 years to keep the nation on a 30-year path to net zero, its findings are also relevant to any deep decarbonization effort that would substantially reduce emissions over more than 10 years. *Notably, the task statement does not pose the question of whether climate impacts of fossil emissions justify deep decarbonization, but rather charges the committee to analyze and understand alternative decarbonization pathways.*

The statement of task also does not ask how deep decarbonization in the United States fits into a broader climate policy including adaptation, global cooperation, and perhaps solar geoengineering. An effective climate policy will need to contemplate all of these components, ensuring an appropriate and effective mix, particularly as information, action, and demands evolve over time.

The statement of task calls for interim and final reports. This interim report focuses on the electricity, transportation, industrial, and buildings sectors, which comprise most of the energy system, and CO_2 emissions, the GHG with the greatest climate impact. In what follows, "energy system" is used as a shorthand for the union of the electricity, transportation, industrial, and buildings sectors. The committee understands that reaching net zero will require addressing all emissions sectors and GHGs. Its final report will include agriculture emissions, expanded treatment of technologies (e.g., hydrogen, low-carbon fuels, negative emissions technologies), and policy actors (state, local, private sector, nongovernmental organizations). It will also consider

BOX 1.2
STATEMENT OF TASK

Building off the needs identified at the *Deployment of Deep Decarbonization Technologies* workshop in July 2019, the National Academies of Sciences, Engineering, and Medicine will appoint an ad hoc consensus committee to assess the technological, policy, social, and behavioral dimensions to accelerate the decarbonization of the U.S. economy. The focus is on emission reduction and removal of CO_2, which is the largest driver of climate change and the greenhouse gas most intimately integrated into the U.S. economy and way of life. The scope of the study is necessarily broad and takes a systemic, cross-sector approach. The committee will summarize the status of technologies, policies, and societal factors needed for decarbonization and recommend research and policy needs. It will focus its findings and recommendations on near- and midterm (5–20 years) high-value policy improvements and research investments and approaches required to put the United States on a path to achieve long-term net-zero emissions. This consensus study will also provide the foundation for a larger National Academies initiative on deep decarbonization. The committee will produce an interim report and a final report. The interim report will provide an assessment of no-regrets policies, strategies, and research directions that provide benefits across a spectrum of low-carbon futures. The final report will assess a wider spectrum of technological, policy, social, and behavioral dimensions of deep decarbonization and their interactions. Specific questions that will be addressed in the final report include the following:

- *Sectoral interactions and systems impacts*—How do changes in one sector (e.g., transportation) impact other sectors (e.g., electric power) and what positive and negative systems-level impacts arise through these interactions; and how should the understanding of sectoral interactions impact choices related to technologies and policies?
- *Technology research, development, and deployment at scale*—What are the technological challenges and opportunities for achieving deep decarbonization, including in challenging activities like air travel and heavy industry; what research, development, and demonstration efforts can accelerate the technologies; how can financing and capital effectively support decarbonization; and what are key metrics for tracking progress in deployment and scale up of technologies and key measurements for tracking emissions?
- *Social, institutional, and behavioral dimensions*—What are the societal, institutional, behavioral, and equity drivers and implications of deep decarbonization; how do the impacts of deep decarbonization differ across states, regions, and urban versus rural areas and how can equity issues be identified and the uneven distribution of impacts be addressed; what is the role of the private sector in achieving emissions reductions, including companies' influence on their external supply chains; what are the economic opportunities associated with deep decarbonization; and what are the workforce and human capital needs?
- *Policy coordination and sequencing at local, state, and federal levels*—What near-term policy developments at local, state, and federal levels are driving decarbonization; how can policies be sequenced to best achieve near-, medium-, and long-term goals; and what synergies exist between mitigation, adaptation, resilience, and economic development?

discussion of wider societal trends, such as changes in economics, demographics, housing patterns, and infectious disease incidents that impact the energy system. During the development of its interim report, the committee discussed issues it sees as important for the final report, although the specific topics and structure of its final report have not been determined.

A complete transformation of the U.S. energy economy would dramatically affect most facets of society and thus have an impact on many areas of national concern, including environmental issues; public and economic health; job losses, gains, and quality; the distribution of income; the treatment of minority and indigenous people; and U.S. international leadership. As a result, net-zero policy is not about energy alone, because a host of other issues that people care deeply about would also be strongly impacted by the way in which net zero is achieved.

The committee studied how alternative policies, all of which could achieve net zero, would differentially affect other national objectives. Its membership was formulated by the National Academies to encompass a diversity of perspectives and expertise, including expertise in economics, the natural sciences, energy technology, political science, public policy, the social dimensions of technological change, labor, geography, and environmental justice. The portfolio of highest-priority policies in this report reflects this diversity of perspectives, because it attempts to find balance between alternative value propositions.

The remainder of this chapter offers four different, but not mutually exclusive, lenses that the committee brought to the net-zero problem, followed by a road map to Chapters 2 through 4. The first emphasizes cost minimization, the second equity and social justice, the third the enhanced competitive position of the United States in a net-zero world because of the country's unique natural resources, and the fourth the opportunity to rebuild the industrial sector of our economy and enhance job quality while maintaining technological leadership. The committee views all of these lenses as critical to attain a robust and sustainable energy transition. The key is to formulate a policy portfolio that balances insights from alternative lenses, rather than to rely too heavily on any single lens.

PERSPECTIVES ON THE NET-ZERO PROBLEM

Economics

All else equal, policy should be formulated to achieve the climate and health benefits of net zero at the lowest possible cost. The classical view from economics is that the transition to net zero will be costly, and justified if the impacts avoided by reduced climate change and fossil pollution outweigh added costs associated with the net-zero system. In addition to climate change, fossil emissions are responsible for the majority of air pollution, which kills millions every year globally. Annual deaths linked to fossil fuels in the United States alone have been estimated as high as 200,000 (Caizzo et al., 2013; Lelieveld et al., 2019). There are many other references on this potential co-benefit to decarbonizing the U.S. economy (e.g., Prehoda and Pearce, 2017; Dimanchev et al., 2019; Patz et al., 2020), and for other countries with extreme air quality problems, this co-benefit easily overwhelms climate benefits at least in the short term (Markandya et al., 2019). A net-zero energy system in the United States would prevent most deaths linked to fossil fuels and provide other health and environmental benefits.

The United States cannot solve the global climate problem on its own because it is responsible for only 10 percent of current emissions. The United States is, however, after China, the second largest emitter, and the largest historical emitter (Friedlingstein et al., 2019). The climate benefit of a U.S. transition to net zero is thus twofold: (1) reducing a significant share of global GHG emissions, and (2) encouraging others to do the same by driving down technology costs and leading a global coalition of nations that collectively make the transition. As noted above, there are enormous co-benefits from decarbonizing the U.S. energy system and economic opportunities for U.S. companies that lead this effort. Ultimately, all of these climate and nonclimate benefits and costs could be combined in an analysis of the net-zero goal. Such an approach would focus heavily on the social cost of carbon (EPA, 2015; NASEM, 2017), which describes, in monetary terms, the harms caused by a marginal ton of CO_2e GHG emissions. The committee notes that such measures necessarily ignore some consequences that are difficult to monetize.

However, the committee was tasked to evaluate paths to net zero, not to decide whether a transition to net zero is justified. For this reason, "cost-effectiveness" is a more relevant economic metric than benefits minus costs. The cost-effectiveness of a policy measures how much the policy costs to achieve a given objective—in this case, in terms of what households or the government must give up, compared to the least-cost alternative that achieves the established objective (here, net-zero emissions). An economy-wide price on carbon tends to be the most cost-effective option in this narrow sense, but cannot by itself address a host of important issues that will inevitably arise, including the need to protect historically disadvantaged communities, communities adversely affected by the energy transition, and U.S. manufacturing that competes in a global transition. For example, if the United States begins the transition before some of its economic competitors without such protections in place, both domestic manufacturing and CO_2 emissions may simply shift overseas.

Among the specific considerations not addressed by a carbon price, many relate to uncertainties. For example, government intervention may be needed in private capital markets, because essential net-zero investments early in the transition may be viewed as too risky, given uncertainties about whether the government will not follow through with the policies that would make the investments profitable. This is especially true for infrastructure. Performance standards may be required in some sectors because people often are uncertain whether they will realize a net economic gain from more efficient equipment, especially when retrofitting their homes or replacing their appliances or vehicles with those that require higher up-front costs and will provide efficiency benefits only in the future.

Cost-effectiveness analysis also ignores both benefits and, typically, how costs and benefits are distributed within an economy. Separate policies, including choices about how to use the revenue from carbon pricing, will thus be needed to meet any distributional objectives, such as those discussed below.

Last, a high carbon price would likely be required to drive the economy to net-zero emissions using carbon pricing alone. Based on existing studies, it is unclear whether competitiveness and equity concerns can be convincingly addressed at such high prices. Therefore, the committee chose to limit the carbon price and turn to other policies, with some loss of cost-effectiveness, in order to manage these concerns.

Equity and Fairness

The transition to net zero provides a unique opportunity to build an energy system that is fair to all Americans and to help redress past discrimination and build a more just society. The committee adopts a broad definition of equity and fairness in the distribution of benefits, costs, impacts, burdens, opportunities, participation, and outcomes associated with the transition to net-zero carbon emissions in the energy system. The committee is concerned both about leveraging the transition to net zero to make energy systems fairer and to reduce historical injustices as well as about ensuring that the transition itself treats all Americans fairly and equitably. Equity also includes the potential for targeted restorative investment strategies in disadvantaged communities, including but not limited to those that have confronted undue burdens associated with current or historical energy systems. The current U.S. energy system unfairly burdens low-income and BIPOC (Black, Indigenous, people of color) households and communities. These communities have disproportionately large exposure to pollution from energy infrastructure, but receive a disproportionately small share of energy revenues, and have comparatively little say in decision making that shapes local energy services and infrastructure (Hajat et al., 2015; Mikati et al., 2018). Low-income and minority communities often have undependable energy services, with frequent outages (Hernández and Laird, 2019). Energy costs take a disproportionately large fraction of low incomes, which leads to a cycle of energy poverty (Drehobl and Ross, 2016; Lyubich, 2020). Location also has an impact; rural communities, in addition to having lower average income, often bear larger energy burdens than their suburban and urban counterparts (Ross et al., 2018). Any financial setback, such as a medical expense, layoff, insufficient job hours, or disability, can lead to inability to pay for energy and withdrawal of service, which both exacerbates the initial setback and impedes recovery. Moreover, low-income households receive disproportionately low benefits from improved energy technology, and public incentives that promote it, because they often do not own their homes, and if they do, they frequently lack the capital for an upgrade that would pay for itself over time, or do not meet the compliance with code necessary to accomplish an upgrade without incurring additional expenses (Hernández et al., 2016; Jessel et al., 2019). Energy poverty can result in inconsistent energy access and extreme temperatures in the home, which have been connected to negative health effects (Ross et al., 2018). This exacerbates health risks among already vulnerable communities.

A transition to a net-zero energy system is thus an opportunity to build an energy system without the injustices that permeate our current system, and for those that are marginalized today to share equally in any future benefits. Also, every technological transformation eliminates jobs tied to the old technology even as it creates new jobs, and drives critical employers in some communities out of business, while adding new employers in others. Policies during the transition must address injustice and loss simply because, in addition to ethical or religious concerns, significant opposition by any group or region of sufficient size could endanger the entire effort. With appropriate policy mechanisms, disadvantaged communities may see significant co-benefits such as high-quality jobs, economic opportunities, and improvements in air quality.

Because energy use affects so many aspects of people's lives, a three-decade transition to net zero simply cannot be achieved without the development and maintenance of a strong social contract. This includes support for a carbon tax, clean energy standard for electricity, electrification of vehicles and buildings, and the founding of a Green Bank and National Transition Corporation. The United States will need specific policies to cultivate public support for the transition, ensure an equitable and just net-zero energy system, and facilitate the recovery of people and communities hurt by the transition. This is imperative to create and maintain the social contract and accomplish the mission. It would also help redress past injustice and help to build a more just society.

Energy Technology

The United States has a unique set of assets that should allow the country to transition at lower cost than many other nations and provide competitive advantage in a decarbonized world. A net-zero energy system that the United States could build over the next 30 years would have the following five components:

- *Zero-carbon electricity.* Especially when the cost of avoiding CO_2 is taken into account, the United States has several cost-competitive energy sources, including wind, solar, hydro, geothermal, and existing nuclear. New wind and solar now offer the cheapest levelized cost of electricity over most of earth's surface (IRENA, 2020). Operating expenses for existing coal plants are often higher than building and operating the equivalent renewable capacity (Figure 1.1). The levelized costs of wind has declined by 70 percent and solar photovoltaics by almost 90 percent since 2009, providing an important means to supply electricity with no direct CO_2 emissions (Lazard, 2019; LBNL, 2020). Hydropower, energy storage, bioenergy, geothermal, nuclear energy, and

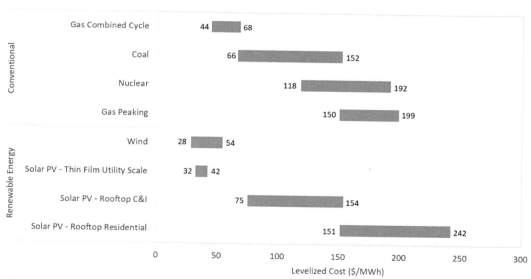

FIGURE 1.1 Selected renewable energy generation technologies are cost-competitive with conventional generation technologies under certain circumstances (e.g., solar can be more expensive than natural gas and coal when installed on rooftops, but is cheaper than both when it is thin film utility scale), and on a levelized cost of energy basis. SOURCE: Adapted from Lazard, Levelized Cost of Energy Analysis, Version 14.0.

natural gas with carbon capture and sequestration are available to compensate for the intermittency of wind and solar electricity.
- *Electrification of transportation and heat in buildings.* Light- and medium-duty vehicles would transition to electric power, while residences would be heated by electric heat pumps. The cost of lithium-ion batteries dropped by 85 percent over the past 10 years (BNEF, 2019).
- *Carbon capture, utilization, and sequestration (CCUS).* To address hard-to-decarbonize sources in industrial and other sectors, CCUS could provide a means to reduce emissions from industrial processes that release CO_2, such as cement production, and perhaps for fossil electric power. There were 51 large-scale carbon capture and sequestration (CCS) operations around the world by 2019, demonstrating at scale virtually all practical applications of CO_2 capture (Global CCS Institute, 2019).
- *Net-zero liquid and gaseous fuels for applications that require high energy density such as airliners or high-temperature industrial process heat.* Options include biofuels, synthetic hydrocarbon fuels, and hydrogen from biomass gasification, electrolysis and natural gas with CCS.
- *Sinks to offset emissions that are too expensive or disruptive to mitigate (i.e., some agricultural methane and N_2O).* Options include forest planting, rebuilding the carbon content of agricultural soils with alternative agricultural practices, and direct air capture (DAC—machines that extract CO_2 from the air), which is still expensive but coming down in price (NASEM, 2019). Although technologically feasible, CCUS coupled to hard-to-decarbonize industries including steel and cement, and net-zero fuels in the quantities needed for aviation, marine transport, fuel-cell heavy trucks, and industrial heat are not yet ready for commercial deployment. If innovation fails to bring any of these to commercial readiness in time, then additional deployment of negative emissions technologies would be needed to offset them.

The past 10 years have seen a revolutionary expansion of cost-competitive energy technologies, unlike anything in the previous 150 years. A transition to net zero would be difficult to contemplate without recent rapid cost declines in core technologies like wind, solar, and electric vehicles (EVs), and the revolution continues. Technological advances are being made in "clean firm" resources, such as advanced modular nuclear reactors, natural gas with CCS, and carbon-free fuels, which can provide a base for dependable electricity that can work in concert with renewables and energy storage to manage demand peaks and weather events. Research and development (R&D) continues to add to the portfolio of options, with new battery chemistries; multiple DAC designs; low-cost designs for electrolysis, which makes hydrogen fuel from water with electricity; and new processes to use captured carbon in products, to name only a few. In this sense, the net-zero movement is as much an outgrowth of technological revolution as it is a response to climate change.

The United States is well positioned for a transition to net zero because of its unique combination of abundant sites for solar and wind, abundant natural gas for use with CCS, best-in-world geologic reservoirs for CO_2 disposal, immense agricultural and forestry sectors producing waste biomass and with 40 million acres already devoted to biofuels feedstock, and a managed forest carbon sink already at 700 million metric tons of CO_2 per year that could be augmented with inexpensive technology already in hand (NASEM, 2019). Because of its unique mix of resources, the United States should be able to decarbonize at lower cost than many other nations, and should have a competitive advantage in a decarbonized world.

Energy Policy

With the right policies to guide it, the transition to net zero would restore U.S. leadership in energy technology, manufacturing, and climate policy, and add high-quality jobs and improved energy access to the U.S. economy. Although the United States still leads the globe in technological innovation,[2] it has not capitalized on its traditional first-mover advantage to sustain leadership in manufacturing and exports for clean energy. This is as true in low-carbon energy technology as it is in other fields such as information technology and artificial intelligence, where firms in Europe and Asia now dominate. For example, the United States was the original leader of the solar energy revolution. Bell Labs investments resulted in the creation of the first solar cell, and strong and steady procurement from the Navy and NASA allowed American solar companies Hoffman Electronics (no longer in business), Automatic Power (now Pharos Marine Automatic Power), and Solar Power Corporation (originally funded by Exxon, which shut it down in the mid-1980s) to serve that market (Nemet, 2006). U.S. labs and companies continue to routinely invent new solar cells that set world records for efficiency in converting sunlight to electricity. In wind, Scottish inventor James Blyth created the first electricity-generating wind turbine in 1886 while serving as a professor at Anderson's College (now the University of Strathclyde—a leader in offshore wind research). American inventor Charles Brush of Cleveland, Ohio,

---

[2] According to the National Science Foundation (NSF) 2020 State of U.S. Science and Engineering report, "The United States continues to perform the largest share of global research and development (R&D), generate the largest share of R&D-intensive industry output globally, award the largest number of S&E doctoral degrees, and account for significant shares of S&E research articles and citations worldwide."

also constructed a homemade wind turbine in his backyard, shortly after Blyth. Brush Electric Company was eventually bought by what is now General Electric (Owens, 2019). In electric vehicles, Tesla is the world's top global producer, but it is the only American firm in the top eight producers (four are Chinese, one is Japanese, one Korean, and one French).

Since the turn of the century, however, the United States has ceded much of its original leadership in these low-carbon industries. Only one of the top-10 solar photovoltaic (PV) manufacturers, First Solar, is an American firm (eight are Chinese, one is South Korean), and U.S. companies' share of the global solar market has dropped below 10 percent (Sonnischen, 2020). Of the top-5 lithium-ion battery producers, there is only one American firm, Tesla, and it ranks fifth behind Korean, Chinese, and Japanese producers (Benchmark Mineral Intelligence, 2019). In 2005, Denmark had the world's largest wind turbine manufacturing capacity, closely followed by Germany and the United States. Yet, in only 15 years, China surged to become the largest manufacturer of wind turbines globally, with six times the U.S. manufacturing capacity. Denmark leads the world in wind power equipment exports, followed by Germany. Wind power equipment exports from the United States are significantly lower.

These trends are disturbing. Manufacturing is important to the U.S. economy, creates high-wage and high-skill jobs, and has a vital impact on innovation and competitiveness. The industrial sector is essential to produce the materials, components, and technology necessary for modern life. The United States should attempt to claw these industrial sectors and markets back, so that it leads the world both in innovation and in the manufacturing and commercialization of advanced clean energy technologies. Surging from behind to win the race will require an integrated national strategy, involving a mix of innovation and smart industrial policy (see Chapter 4 for details) that positions U.S. firms to compete in the highly competitive international landscape in clean energy.

A major reason why the United States has not maintained its competitiveness in clean energy industries is the inconsistency and unpredictability of market-formation policies in its domestic market and weak export promotion policies to help U.S. firms succeed in the global marketplace relative to other countries (Lewis and Wiser, 2007; Gallagher, 2014). The U.S. production tax credit for wind has been extended 12 times since it was enacted in 1992, and in 7 of those cases the credit expired before it was retroactively extended (CRS, 2020). The United States never passed a national clean energy standard (although Renewable Portfolio Standards exist in a majority of the states). It also never created a feed-in tariff for clean energy, unlike Germany, China, and Japan. In R&D investments, volatility in appropriations for the energy efficiency and renewable energy programs at the U.S. Department of Energy (DOE) contributed to a lack of certainty about future funds to support innovation (Gallagher and Anadon, 2020), despite strong evidence that investments in energy research, development, and demonstration (RD&D) provide substantial financial returns (NRC, 2001; Wiser et al., 2020). The U.S. Export-Import Bank stopped lending altogether for a period in 2015 and suffered a series of starts and stops in the reauthorization of its charter in the subsequent few years.

As the U.S. clean energy economy continues to grow rapidly, a key consideration will be to ensure that U.S. workers and businesses benefit significantly and that the United States maintains a strong workforce in the energy economy. As the history of the U.S. automobile, computer, information, data analytic, and digital communications industries have demonstrated, continuous innovation both within existing technology domains and in disruptive technologies is key to long-term economic prosperity and the prospects for high-skill, high-wage jobs. Such jobs are necessary to create a robust foundation for both the U.S. economy as a whole and the economic security of individuals, households, and communities. Yet, as the decline of the U.S. automobile industry across the upper midwestern United States has illustrated since the 1980s, and recent trajectories in the gig economy in the information technologies sector also demonstrate, U.S. policies have not always managed the risks of disruptive innovation well.

As decarbonization expands, therefore, it will be important for U.S. policy to attend carefully to both the risks of significant declines in carbon-based energy industry workforces and businesses (e.g., gasoline sales and internal combustion engine parts and repair) and the need to ensure that U.S. clean energy jobs are high quality. A high-quality job entails, at a minimum, a safe and secure working environment, family-sustaining wages[3] and comprehensive benefits, regular schedules and hours, and skills-development opportunities that enable wage advancement and career development (United Way Worldwide, 2012; AFL-CIO, 2017; ILO, 2020). The United States will also need robust educational and workforce training and development programs for the clean energy sector across a wide array of diverse technology and business domains.

---

[3] A family-sustaining wage is how much wage-earning individuals in a household must earn to support themselves and their family, working full time (Glasmeier and MIT, 2020a,b). Some examples: In North Carolina, which sits in the middle of state rankings for cost of living, two working adults in a household with two children would need to be paid at least $15.85/hr each. If they lived in the D.C.-Arlington area, known for its high cost of living, a family-sustaining wage would be $18.06/hr each.

Ultimately, the goal of decarbonization policy should be to develop a comprehensive, integrated approach to a clean energy transition that ensures that the U.S. energy workforce becomes larger, better compensated, and more secure than it is today.

ROAD MAP TO THE REST OF THE REPORT

The rest of the report is organized around a series of questions: (1) What is needed from a technological point of view to reach net zero? (2) What other goals besides GHG emissions reductions should guide the transition? (3) What suite of policies is needed in the first 10 years to embark on a transition to net zero?

Chapter 2 addresses the first question, reviews the literature on paths to net zero, and concludes that net zero by 2050 is achievable technically and economically—that is, such outcomes are potentially achievable at roughly the same level of spending (approximately 4 percent of gross domestic product [GDP]) that the nation expends on energy services today (Larson et al., 2020). In the committee's analysis, a change in mindset is required by those who have spent years focused on the least expensive way to reduce carbon emissions on the margin in a short-term economic sense. In the committee's view, achieving a 30-year transition to net zero at the lowest cost means investing in some of the higher marginal cost projects up-front, to take advantage of the natural turnover of long-lived capital stock, and to facilitate later phases of the transition (i.e., retrofitting power plants even if it would be immediately cheaper per ton of emissions avoided to plant trees).

Chapter 2 identifies five actions that would need to be taken in the 2020s to put a net-zero energy system within reach by 2050. These five actions represent islands of relative certainty, because any plan to achieve net zero at midcentury is constrained by the immediate need to replace long-lived emitting components as they retire and to meet any expansions in demand with non-emitting assets, and because any large-scale deployment over the next decade must necessarily rely on proven, mature technologies. Also, the list of actions recommended by the committee for the 2020s is relevant to the final make-up of the energy system in 2050. These actions would all be needed regardless of whether the final system is to be 100 percent renewable or retains substantial nuclear and non-emitting fossil fuel components. Last, the five recommended actions are also robust to uncertainty caused by a future technological breakthrough, such as low-cost DAC or electrolysis. The 30-year time horizon means that the United States cannot wait until a new breakthrough occurs (if ever), especially given that any new innovation would take years or even decades to bring to material scale. These actions are therefore designed to make immediate and necessary progress, to lay the foundations to reach net zero by 2050, and to retain optionality to manage risk and uncertainty in the later portion of the transition.

The five required actions are:

1. *Electrify energy services in transportation, buildings, and industrial sectors.* Examples include, by 2030, reaching half of vehicle sales (all classes combined) from zero-emissions vehicles (electric and fuel cell), and deploying heat pumps in one-quarter of residences.
2. *Improve efficiency and energy productivity in transportation, building, and industrial sectors.* There are many examples of low-hanging fruit in this category, including improved efficiency of appliances and buildings, and accelerating the rate of increase of industrial energy productivity (dollars of economic output per energy consumed) from recent rates of 1 percent per year to 3 percent per year (Morrow et al., 2017).
3. *Carbon-free electricity.* Roughly double the share of electricity generated by carbon-free sources from 37 percent to about 75 percent by 2030, including deployment on the order of 600 GW of wind and solar power capacity.
4. *Build critical infrastructure needed for the transition to net zero.* Examples include substantial expansion of high-voltage transmission lines to move renewable power between regions, a national CO_2 transportation network to move captured CO_2 to geologic reservoirs (useful for decarbonizing industry and producing carbon negative fuels even in a 100 percent renewable system), and an expanded network of EV charging stations.
5. *Expand the innovation toolkit.* Examples include RD&D for electrolysis to make fuels from renewable power, inexpensive DAC, which could be used to offset any GHG emissions that prove to be too difficult or disruptive to mitigate, and any innovation that would further reduce the cost of technologies that are already cost-effective.

These five actions would put the nation on a path to a net-zero energy system able to meet the nation's projected business-as-usual demand for energy services, and would not require dramatic reductions in service demand, such as significantly reduced mobility or home size. The goals include significant increases in energy efficiency through electrification of transport and heating and changes to buildings and industry, which would reduce the demand for energy rather than the demand for energy services. The committee was not confident in its ability to design policy that would both attract public support and achieve the behavioral changes required for a significant reduction in the demand for energy services.

Complementary to the five critical actions, Chapter 2 describes decarbonization strategies by sector, providing requirements for buildings, transportation, industry, energy storage, fuels, electricity generation and transmission, and CCS. In addition to addressing these actions to decarbonize the U.S. energy system, the United States must also tackle non-CO_2 GHGs and preserve and enhance land carbon sinks. Although the statement of task focuses on CO_2, the committee briefly summarizes actions required to reduce methane, N_2O, and fluorinated gas emissions in the three end-use sectors and to offset remaining emissions of these gases with forestry and agricultural carbon sinks in the Addendum on Non-CO_2 Greenhouse Gases and in Box 2.1, both in Chapter 2. The final report will address the forestry and agricultural policies required to produce and sustain the needed CO_2 sinks.

Chapter 3 most clearly distinguishes this report from others that characterize technological pathways. It develops four socioeconomic goals that address critical issues of national concern that are implicated in a net-zero transition:

1. *Strengthen the U.S. economy.* Provide the nation with reliable, low-cost, net-zero energy, while using the transition to accelerate U.S. innovation, reestablish U.S. manufacturing, increase the nation's global economic competitiveness, and increase the availability of high-quality jobs.
2. *Promote equity and inclusion.* Benefits, risks, and costs of the transition to net zero should be equitably distributed. Historically marginalized groups should be fully integrated into decision making.
3. *Proactively support workers, businesses, and communities directly and adversely affected by the transition.* Promote fair access to new long-term employment opportunities and provide financial and other support to communities that might otherwise be harmed by the transition.
4. *Maximize cost-effectiveness.* Cost-effectiveness measures the material consumption given up by households in order to achieve net zero in 2050, relative to a business-as-usual counterfactual.

There are two issues of national concern that the committee did not explicitly address when evaluating net-zero policies. The first is COVID-19. The COVID-19 pandemic has affected many aspects of everyday life in 2020 and could have significant impacts on short- and long-term economic conditions and decarbonization initiatives. Ongoing and projected behavioral changes, including shifts in transportation modes (away from public transportation and toward personal vehicles, walking, or cycling), increases in telework and online purchasing, and relocation outside urban centers all influence the opportunities and strategies for a net-zero energy transition (IEA, 2020a). The decreases in travel, industrial and trade activities, and demand for electricity and oil in 2020 have reduced global CO_2 emissions by about 4 to 11 percent relative to 2019 levels (IEA, 2020a; Climate Action Tracker, 2020). At the same time, however, the economic fallout from the pandemic has decreased investment in and development of renewable, clean, and energy-efficient technologies, at least in the short term (IEA, 2020b). The long-term effects of these actions on future emissions reductions remain uncertain. Nonetheless, there is general agreement that economic recovery packages designed to promote clean energy policies and investments are critical for achieving deep decarbonization and also provide opportunities to increase equity and sustainability (IEA, 2020a; Climate Action Tracker, 2020). However, the committee's recommendations focus on longer-term policies.

The second issue is related to national security, including managing materials sources and intellectual property to increasing manufacturing capabilities and training the workforce. There are also obvious national security implications of a global switch to net zero, but the committee did not include experts on national security to address these considerations. Further, climate change itself has critical national security consequences. Even a 1 to 2 degrees Celsius warming would result in more intense and frequent natural disaster events, with significant losses of life and property, and greater spending by the federal government on responding to such disasters (Guy et al., 2020; Kaplan, 2020). Impacts to military installations from severe weather, river flooding, hurricanes, and extreme rain have already cost the U.S. military $10 billion in recent years (Underwood, 2020). The Department of Defense (DoD) characterized climate change in 2014 as a "threat multiplier," meaning that its impacts will amplify stressors like poverty, environmental degradation, political instability, and social tensions (La Shier and Stanish, 2019). With its global presence, the U.S. military will need tailored responses to climate change in each of its geographic regions, including addressing potential destabilizing events stemming from increased drought, disaster, and disease.

In addition, this interim report does not include policies needed to sustain forestry and agricultural carbon sinks to offset emissions that remain too expensive or disruptive to mitigate, including some agricultural emissions of methane and N_2O (see Box 2.1). All anthropogenic negative emissions are technically emissions offsets, and substantial negative emissions will be essential to achieve net zero in 2050. Fortunately, the United States has the required capacity to offset residual emissions of non-CO_2 GHGs in its forestry and agricultural sectors, and the economy-wide price on carbon proposed in Chapter 4 should be sufficient to sustain needed agricultural and forestry sinks through 2050 (NASEM, 2019, Box 2.1). Although the nation already possesses a land use CO_2 sink of 700 $MtCO_2$/y, additional policies will be needed because the sink is expected to halve by 2050 without deliberate actions to sustain it, and because policy must avoid incentivizing harmful land use change that could damage the nation's biodiversity or production of food and fiber. These policies must also prohibit or discourage carbon credits from being used to prevent replacement of long-lived capital stock with non-emitting alternatives (e.g., a new fossil power plant with forestry offsets versus a new plant with carbon capture and sequestration), because this would increase both the total cost of the transition and the amount of sink required to complete it, given that the total sink capacity is limited (NASEM, 2019). The committee decided to defer discussion of the policies to create and manage agricultural and forestry carbon sinks to the final report, because of the complexity of the issues involved, and because the current slowly changing carbon sink will be sufficient for the near term.

Chapter 4 evaluates policies *at the federal level* that the nation could adopt to achieve the five technological actions in Chapter 2 while advancing the socioeconomic goals in Chapter 3. Local, state, and regional policies will be included in the final report. Collectively, the recommended federal policies would catalyze the first 10 years of a transition to net zero, and provide the associated environmental, health, and societal benefits, while controlling costs, protecting the competitiveness of the U.S. economy, and compensating for market failures. They would also increase the number of high-quality manufacturing jobs, while protecting vulnerable workers and communities, and would reestablish U.S. leadership in energy innovation, manufacturing, and commercialization, while building a more just energy system.

For each policy, the committee identified a responsible branch of government and the needed congressional appropriation, if any. The list of high-priority policies is relatively granular (summarized in Table 4.1 in Chapter 4) and is divided into four categories:

1. *Policies to establish a U.S. commitment to a rapid, just, and equitable transition to a net-zero greenhouse gas emissions economy.* A partial list includes the adoption of a national GHG emissions budget; an economy-wide price on GHG emissions; a federal effort to monitor and evaluate equity impacts of net-zero policies; and a National Transition Corporation to mitigate job losses and ensure equitable access to economic opportunities during the transition.
2. *National rules and standards to accelerate the formation of markets for clean energy that work for all.* A representative subset includes standards for the pace of transition to zero-emissions vehicles; manufacturing standards for net-zero appliances; a clean electricity standard for electric power generation; buy American rules, buy clean rules, and labor standards for federal agencies and companies that receive federal funds; changes in electricity wholesale

market rules; and disclosure rules for climate and net-zero policy-related risks covering private companies and federal agencies.
3. *Investments in research, technology, people, and infrastructure needed for the transition to net zero.* A partial list includes a tripling of the nation's RD&D budget for clean energy; a Green Investment Bank; regulatory reform and incentives required to augment the nation's electrical transmission network, particularly over long distances; a national CO_2 transportation network, with characterization and permitting of geologic storage reservoirs; an interstate EV charging network; upgrades in the electric grid; a comprehensive education and training program ranging from the vocational to the doctoral level to prepare the needed workforce; and incentives and loan guarantees to revitalize U.S. clean energy manufacturing, which are tied to labor standards and equity and inclusion goals.
4. *Policies to support coordinated planning for the transition, with effective inclusion of diverse participants.* A subset includes a national interagency working group to facilitate and coordinate the work of all federal agencies on a just transition; 10 regional centers to plan the transition at the regional level, an office in each state to coordinate federal and state action; community-based demonstration projects for programs designed to strengthen equity outcomes, and local community block grants for transition planning and to identify communities at risk, with funding tied to effective participation by historically marginalized populations.

REFERENCES

AFL-CIO. 2017. "Resolution 1: Workers' Bill of Rights | AFL-CIO." AFL-CIO. October 25. https://aflcio.org/resolutions/resolution-1-workers-bill-rights.

Benchmark Mineral Intelligence. 2019. "Who Is Winning the Global Lithium-Ion Battery Arms Race?" https://www.benchmarkminerals.com/who-is-winning-the-global-lithium-ion-battery-arms-race/.

BNEF (Bloomberg New Energy Finance). 2019. "Battery Pack Prices Fall As Market Ramps Up With Market Average At $156/kWh In 2019." https://about.bnef.com/blog/battery-pack-prices-fall-as-market-ramps-up-with-market-average-at-156-kwh-in-2019/#:~:text=BNEF's%202019%20Battery%20Price%20Survey,with%20internal%20combustion%20engine%20vehicles.

Caizzo, F., A. Ashok, I. Waitz, S.H.L. Lim, and R.H. Barrett. 2013. Air pollution and early deaths in the United States. Part I: Quantifying the impact of major sectors in 2005. *Atmospheric Environment* 79: 198–208.

CDP. 2019. "The A List 2019." https://www.cdp.net/en/companies/companies-scores.

Climate Action Tracker. 2020. "Pandemic Recovery: Positive Intentions vs Policy Rollbacks, with Just a Hint of Green." https://climateactiontracker.org/documents/790/CAT_2020-09-23_Briefing_GlobalUpdate_Sept2020.pdf.

CRS (Congressional Research Service). 2020. *The Renewable Electricity Production Tax Credit: In Brief.* CRS Report R43453. Washington, DC.

Dimanchev, E.G., S. Paltsev, M. Yuan, D. Rothenberg, C. W. Tessum, J. D. Marshall, and N.E. Selin. 2019. Health co-benefits of sub-national renewable energy policy in the US. *Environmental Research Letters* 14(8).

Drehobl, A., and L. Ross. 2016. *Lifting the High Energy Burden in America's Largest Cities: How Energy Efficiency Can Improve Low-Income and Underserved Communities*. Washington DC: American Council for an Energy-Efficient Economy.

EPA (Environmental Protection Agency). 2015. *Social Cost of Carbon Factsheet*. Washington DC: EPA Archive.

EPA. 2020. *Inventory of U.S. Greenhouse Gas Emissions and Emissions and Sinks 1990–2018*. Washington DC.

Friedlingstein, P., M. Jones, M. O'Sullivan, R. Andrew, J. Hauck, G. Peters, W. Peters, et al. 2019. Global Carbon Budget 2019. *Earth System Science Data* 11: 1783–1838.

Gallagher, K.S. 2014. *The Globalization of Clean Energy Technology: Lessons from China (Urban and Industrial Environments)*. Cambridge, MA: MIT Press.

Gallagher, K.S., and L.D. Anadon. 2020. "DOE Budget Authority for Energy Research, Development, and Demonstration Database." Fletcher School of Law and Diplomacy, Tufts University; Department of Land Economy, Center for Environment, Energy and Natural Resource Governance (C-EENRG); University of Cambridge; and Belfer Center for Science and International Affairs, Harvard Kennedy School. July 8. https://www.belfercenter.org/publication/database-us-department-energy-doe-budgets-energy-research-development-demonstration-1.

Glasmeier, A.K., and MIT (Massachusetts Institute of Technology). 2020a. "Living Wage Calculation for Washington-Arlington-Alexandria, DC." https://livingwage.mit.edu/metros/47900.

Glasmeier, A.K., and MIT. 2020b. "Living Wage Calculation for North Carolina." https://livingwage.mit.edu/states/37.

Global CCS Institute. 2019. *Global Status of CCS 2019*. https://www.globalccsinstitute.com/resources/global-status-report/.

Guy, K., et al. 2020. "A Security Threat Assessment of Global Climate Change: How Likely Warming Scenarios Indicate a Catastrophic Security Future." Center for Climate and Security. Washington, DC: The Council on Strategic Risks. https://climateandsecurity.org/wp-content/uploads/2020/03/a-security-threat-assessment-of-climate-change.pdf.

Hajat, A., C. Hsia, and M.S. O'Neill. 2015. Socioeconomic disparities and air pollution exposure: A global review. *Current Environmental Health Reports* 2: 440–450.

Hernández, D., and J. Laird. 2019. *Disconnected: Estimating the National Prevalence of Utility Disconnections and Related Coping Strategies*. Philadelphia, PA: American Public Health Association.

Hernández, D., D. Phillips, and E.L. Siegel. 2016. Exploring the housing and household energy pathways to stress: A mixed methods study. *International Journal of Environmental Research and Public Health* 13: 916.

IEA (International Energy Agency). 2020a. "Changes in transport behavior during the Covid-19 crisis." https://www.iea.org/articles/changes-in-transport-behaviour-during-the-covid-19-crisis.

IEA. 2020b. "The Impact of the Covid-19 Crisis on Clean Energy Progress." https://www.iea.org/articles/the-impact-of-the-covid-19-crisis-on-clean-energy-progress.

International Labor Organization. 2020. "Decent Work." International Labor Organization. 2020. https://www.ilo.org/global/topics/decent-work/lang—en/index.htm.

IPCC (Intergovernmental Panel on Climate Change). 2018. *Global Warming of 1.5°C*. Geneva, Switzerland: Intergovernmental Panel on Climate Change.

IRENA (International Renewable Energy Agency). 2020. *Renewable Power Generation Costs in 2019*. Abu Dhabi: International Renewable Energy Agency.

Jessel, S., S. Sawyer, and D. Hernández. 2019. Energy, poverty, and health in climate change: A comprehensive review of emerging literature. *Frontiers of Public Health* 7: 357.

Kaplan. S. 2020. "The Undeniable Link Between Natural Disasters and Climate Change." *The Washington Post*. https://www.washingtonpost.com/climate-solutions/2020/10/22/climate-curious-disasters-climate-change/.

Larson, E., C. Greig, J. Jenkins, E. Mayfield, A. Pascale, C. Zhang, S. Pacala, et al. 2020. *Net-Zero America by 2050: Potential Pathways, Deployments, and Impacts*. Princeton, NJ: Princeton University.

LaShier, B., and J. Stanish. 2019. The national security impacts of climate change. *Journal of National Security and Law* 10: 27-43.

Lazard. 2019. "Levelized Cost of Energy and Levelized Cost of Storage 2019." https://www.lazard.com/perspective/lcoe2019.

LBNL (Lawrence Berkeley National Laboratory). 2020. "Wind Technologies Market Report." https://emp.lbl.gov/wind-technologies-market-report/.

Lelieveld, J., K. Klingmüller, A. Pozzer, R.T. Burnett, A. Haines, and V. Ramanathan. 2019. Effects of fossil fuel and total anthropogenic emission removal on public health and climate. *Proceedings of the National Academy of Sciences* 116(15): 7192–7197.

Lewis, J., and R. Wiser. 2007. Fostering a renewable energy technology industry: An international comparison of wind industry policy support mechanisms. *Energy Policy* 35: 1844–1857.

Lyubich, E. 2020. "The Race Gap in Residential Energy Expenditures." Working Paper WP 306. Energy Institute at Haas, https://haas.berkeley.edu/wp-content/uploads/WP306.pdf.

Markandya, A., J. Sampedro, S.J. Smith, R.V. Dingenen, C. Pizarro-Irizar, I. Arto, and M. González-Eguino. 2018. Health co-benefits from air pollution and mitigation costs of the Paris Agreement: A modelling study. *Lancet Planet Health* 2(3): 126–133.

Mikati, I., A.F. Benson, T.J. Luben, J.D. Sacks, and J. Richmond-Bryant. 2018. Disparities in distribution of particulate matter emission sources by race and poverty status. *American Journal of Public Health* 108: 480–485.

Morrow, R., A. Carpenter, J. Cresko, S. Das, D. Graziano, R. Hanes, S. Supekar, et al. 2017. *U.S. Industrial Sector Energy Productivity Improvement Pathways*. 2017 ACEEE Summer Study on Energy Efficiency in Industry. http://aceee.org.

NASEM (National Academies of Sciences, Engineering, and Medicine). 2017. *Valuing Climate Damages: Updating Estimation of the Social Cost of Carbon Dioxide*. Washington, DC: The National Academies Press.

NASEM. 2019. *Negative Emissions Technologies and Reliable Sequestration: A Research Agenda*. Washington, DC: The National Academies Press.

Nemet, G. 2006. Beyond the learning curve: Factors influencing cost reductions in photovoltaics. *Energy Policy* 34(17): 3218–3232.

NRC (National Research Council). 2001. *Energy Research at DOE: Was It Worth It? Energy Efficiency and Fossil Energy Research 1978 to 2000*. Washington, DC: The National Academies Press.

Osko, I., S. Hamburg, D.J. Jacob, D.W. Keith, N.O. Keohane, M. Oppenheimer, J.D. Roy-Mayhew, et al. 2017. Unmask temporal trade-offs in climate policy debates. *Science* 356: 492–493.

Owens, B.N. 2019. *The Wind Power Story: A Century of Innovation That Reshaped the Global Energy Landscape*. Piscataway, NJ: IEEE Press.

Patz, J.A., V.J. Stull, and V.S. Limaye. 2020. A low-carbon future could improve global health and achieve economic benefits. *Journal of the American Medical Association* 323(13): 1247–1248.

Pew Research Center. 2020. *Two-Thirds of Americans Think Government Should Do More on Climate*. Washington DC.

Prehoda, E., and J. Pearce. 2017. Potential lives saved by replacing coal with solar photovoltaic electricity production in the U.S. *Renewable and Sustainable Energy Reviews* 80: 710–715.

Ross, L., A. Drehobl, and B. Stickles. 2018. *The High Cost of Energy in Rural America: Household Energy Burdens and Opportunities for Energy Efficiency*. Washington, DC: American Council for an Energy-Efficient Economy.

Saunois, M., A. Stavert, B. Poulter, P. Bousquet, J. Canadell, R. Jackson, P. Raymond, et al. 2020. The Global Methane Budget 2000–2017. *Earth System Science Data* 12: 1–63.

Sonnischen, N. 2020. "Global Market Share of Solar Module Manufacturers 2017." https://www.statista.com/statistics/269812/global-market-share-of-solar-pv-module-manufacturers/.

Underwood. K. 2020. "The National Security Implications of Climate Change." Signal Media. https://www.afcea.org/content/national-security-implications-climate-change.

United Way Worldwide. 2012. "Financial Stability Focus Area: Family-Sustaining Employment." Alexandria, VA: United Way Worldwide. https://unway.3cdn.net/077876d3896dda796a_hjm6btvnb.pdf.

U.S. Climate Alliance. 2020. *Leading the Charge: Working Together to Build and Equitable, Clean, and Prosperous Future*. 2020 Annual Report. http://www.usclimatealliance.org.

We Are Still In. 2020. "Who's In." https://www.wearestillin.com/signatories.

Wiser, R., M. Bolinger, B. Hoen, D. Millstein, J. Rand, G.L. Barbose, N.R. Narghouth, et al. 2020. *Wind Energy Technology Data Update: 2020 Edition*. Berkeley, CA: Lawrence Berkeley National Laboratory.

CHAPTER TWO

Opportunities for Deep Decarbonization in the United States, 2021–2030

INTRODUCTION

Since the industrial revolution, U.S. greenhouse gas (GHG) emissions have risen steadily in most years, in tandem with an economy fueled by fossil fuels. In recent years, however, the correlation between U.S. economic growth and emissions has weakened. After peaking in 2007, emissions have declined in 7 of the past 11 years, falling 11 percent from 2007 to 2018 (EPA, 2020) even as the economy grew by 19 percent over the same time (OMB, 2020). Nonetheless, emissions are not declining in all economic sectors, and the transition to a zero-carbon economy is not occurring fast enough to meet climate targets.

As discussed in Chapter 1, the United States emits about 6.7 billion metric tons of carbon dioxide equivalent (Gt CO_2e) each year, of which roughly 80 percent is carbon dioxide (CO_2), with the remainder split between methane (10 percent), nitrous oxide (7 percent), and the fluorinated gases (F-gases) (3 percent). Positive changes in land use and forestry offset about 700 million metric tons of carbon dioxide annually, with the result that net U.S. GHG emissions have hovered around 6 billion metric tons of CO_2e over the past several years (2018 data, from EPA, 2020).

As shown in Figure 2.1, when all GHG emissions, including from electricity generation, are distributed by end-use sector, buildings account for the largest share of gross emissions at 32 percent, followed by industry (29 percent), transportation (28 percent), and agriculture (10 percent). When electricity emissions are considered separately, transportation is the top source of direct emissions (28 percent), followed by the electric power sector (27 percent), industry (22 percent), commercial and residential buildings (12 percent), and agriculture (10 percent) (EPA, 2020).

Electric power generation has been the real workhorse of emissions reductions, with carbon dioxide emissions from electricity generation declining by a third from 2005 to 2019 (EIA, 2020a). This decrease resulted from the replacement of the oldest, least-efficient coal plants with output at plants that burn natural gas (up 15 percentage

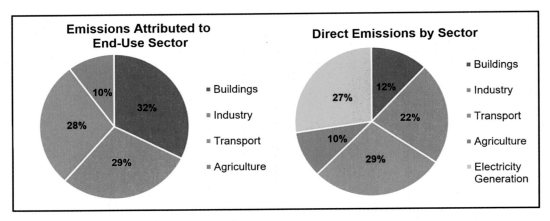

FIGURE 2.1 U.S. gross greenhouse gas emissions in 2018 by sector. The remaining 1 percent of emissions come from U.S. territories, and EPA does not disaggregate these into sectors. SOURCE: Data from EPA (2020).

points of U.S. market share from 2009), and renewable energy, primarily variable generation from wind (up 5.5 percentage points) and solar (up 2 percentage points) (EIA, 2020b). Rapid declines in power sector emissions have been facilitated by the low cost of extracting natural gas from shale formations and precipitous declines in the cost of new solar photovoltaics (PVs; 89 percent cheaper since 2009) and new wind facilities (70 percent cheaper since 2009) (Lazard, 2019). All three of these trends have been driven by proactive public policy support, although these technologies were nascent and still costly (Trembath et al., 2012; Cox et al., 2015; Nemet, 2019; DOE-EERE, 2020b).

Thanks to ongoing policy support and steady innovation by the private sector (and preferences among many corporations for renewable power), the electricity sector could deliver as much as 90 percent clean electricity by 2035 at rates comparable to today's levels. Such an outcome could occur by retaining existing hydropower and nuclear capacity, accelerating deployment of wind and solar to displace coal and some gas-fired generation, retaining most existing natural gas power plants for reliability and flexibility purposes, and building out sufficient electric transmission capacity to connect new renewable generation to the grid (Phadke et al., 2020).

However, there are limits to the quantity of cost-effective emissions reductions achievable with mature technologies, even in the power sector. Even taking into consideration the future coal-plant retirements that have already been announced, there could still be significant coal plant capacity online by 2030, unless competitive pressure increases over time (EIA, 2020c). Some of the remaining coal plants are owned by traditional investor-owned and publicly owned utilities, with their coal-plant investment costs included in the utility's rate base and recovered through retail rates, and are therefore partially shielded from market forces. Additionally, some coal plants provide local reliability service and may not be able to retire unless their capacity is replaced in the near term with sufficient amounts of other resources (e.g., new gas-fired capacity) capable of providing such services, and it may be difficult, if not impossible, to get approvals for such new fossil units. Also, some existing nuclear reactors have been unable to recover their costs in competitive wholesale markets, in part because current markets do not value the carbon-free attribute of electricity generated from nuclear plants. This is especially true for single-unit nuclear power plants and those that are not supported by state policies (e.g., New Jersey's Zero Emissions Certificate Law). The retirement of nuclear power plants will need to be offset by additional net-zero carbon generation to continue making forward progress toward decarbonization goals. While natural gas plants can continue to provide reliability and flexibility services in the near term, reaching a 100 percent carbon-free electricity sector will ultimately require deployment of one or more "clean firm" electricity sources, including geothermal energy, biogas, nuclear energy, natural gas with carbon capture and sequestration (CCS), and hydrogen or other carbon-free fuels produced from net-zero carbon processes. Clean firm resources offer the benefit of carbon-free, dispatchable electricity that is available on demand for as long as needed without dependence on weather, and are thus critical complements to weather-dependent variable renewables and energy-constrained electricity storage technologies (Sepulveda et al., 2018).

Emissions from end-use sectors have not declined as rapidly, and in some cases have even increased. Since 2005, direct emissions (i.e., not accounting for electricity consumption) from transportation and industry declined by 5 percent and 2 percent, respectively. Emissions from agriculture and buildings grew by 5 percent and 6 percent (EPA, 2020). Across the end-use sectors, the story has been remarkably consistent: Increased activity in each sector has been partially offset by moderate levels of efficiency improvements, resulting in only incremental changes in emissions. In the transportation sector, growth in vehicle miles traveled has been offset by improved fuel economy. In the industrial sector, increased economic output has been offset by a combination of more efficient industrial processes and structural changes in the economy (e.g., a shift away from energy-intensive manufacturing to the services industry). And in the buildings sector, growth in floor space has been offset by improved efficiencies of buildings and appliances.

Deep decarbonization of the transportation, industry, and buildings sectors will require taking full advantage of a broader suite of decarbonization tools, including (1) accelerating improvement in end-use efficiency to reduce total fuel and materials demand; (2) substituting hydrocarbon fuels with carbon-free electricity; (3) using "drop-in" hydrocarbon fuels with net-zero lifecycle GHG emissions; and

(4) using CCS, enhanced land carbon sinks, or increases in negative emissions technologies (NETs) to capture or offset emissions from residual fossil fuel use.

This interim report focuses on actions to decarbonize the U.S. economy as part of efforts to reduce net GHG emissions—across all gases—to zero by midcentury. Figure 2.2 provides an illustrative path to achieving net-zero emissions, in which gross carbon dioxide emissions from the end-use sectors are almost completely eliminated, and negative emissions technologies are scaled up to offset residual emissions from hard-to-abate energy sectors. Non-CO_2 gases and land sinks are discussed in Box 2.1.

Some end-use subsectors will be difficult (or prohibitively expensive) to decarbonize completely by 2050. In particular, aviation and shipping are more challenging to electrify than other transportation sectors, and low-carbon fuels may not reach sufficient scale by midcentury. Many industrial sectors, such as cement, iron, and steel and chemicals manufacturing, pose unique decarbonization challenges—for example, decarbonization options for high-temperature heat (Friedmann et al., 2019) and industrial process emissions (de Pee et al., 2018; Rissman et al., 2020), and sector-specific integration challenges. While technologies exist to cut emissions in these sectors,

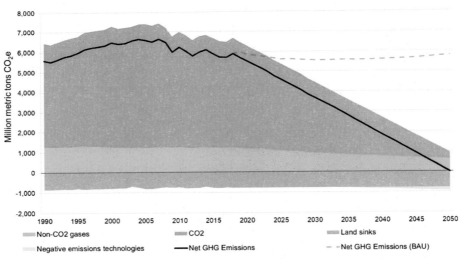

FIGURE 2.2 An illustrative path to net-zero greenhouse gas (GHG) emissions by 2050, by gas. Carbon dioxide (CO_2) emissions from fossil fuel combustion and other sources decline to 5 percent of 2005 levels, with residual emissions from hard-to-abate sectors. In accordance with the IPCC pathways consistent with 1.5°C of warming, methane and nitrous oxide emissions decline to 35 percent and 75 percent of their 2010 levels. The U.S. land sink is maintained at current levels. NETs begin removing atmospheric CO_2 on a large scale in 2035, and scale up to roughly 150 million metric tons annually by 2050. This is one of many possible paths to net-zero emissions and illustrates the key ingredients or building blocks of a net-zero emissions economy: (1) deep reductions in CO_2 emissions (deep decarbonization); (2) declines in non-CO_2 GHGs; (3) maintenance or expansion of land carbon sinks; and (4) expansion of negative emissions technologies.

they remain at precommercial or first-of-a-kind demonstration stages and require significant improvement in cost and performance to become commercially viable. Proactive innovation and maturation of emerging technologies over the next decade could ultimately supply a range of decarbonization options, even in these difficult-to-decarbonize sectors, but the feasibility of complete decarbonization by 2050 remains uncertain. Negative emissions technologies such as direct air capture and storage (DACS) and bioenergy with carbon capture and sequestration (BECCS) may be needed to offset these residual emissions, and provide additional tools in the decarbonization toolkit. To achieve net-zero CO_2 emissions, residual emissions from the energy end-use sectors and negative emissions must sum to zero.

LESSONS FROM DEEP DECARBONIZATION STUDIES AND THE HISTORY OF ENERGY INNOVATION

This report builds on a rich literature of research exploring what a net-zero emissions economy looks like and how to make this transition. Previous deep decarbonization studies vary in their specific technology and policy recommendations, but all share several common core elements. Specifically, the studies promote pathways that combine the following:

- Reducing overall energy demand through increased energy and materials efficiency;
- Decarbonizing electricity generation;
- Switching to electricity and low-carbon fuels in buildings, transportation, and industry (which often involves lower overall energy use in addition to electrification);
- Capturing carbon from residual use of fossil fuels at stationary sources (e.g., fossil power plants, cement, ammonia production);
- Reducing non-CO_2 emissions; and
- Enhancing land sinks and negative emissions technologies to offset all remaining direct emissions.

Most importantly, these analyses find that deep decarbonization is technically feasible at relatively low cost.

1. Deep decarbonization is technically feasible, but proactive innovation is essential.

Deep decarbonization studies find that reaching net-zero emissions is technically feasible (and relatively low cost) provided that significant proactive effort is invested over the next decade to drive the maturation and improvement of a range of more nascent technologies and solutions needed to reach net-zero emissions. For example, the International Energy Agency (IEA) finds that nearly half of the global annual emissions reductions necessary to achieve a net-zero energy system by 2050 will likely have to come from technologies that are currently at the demonstration or prototype stage of development but are not yet commercially available (IEA, 2020a). Although nascent, all of these technologies are technically feasible and do not require fundamental scientific "breakthroughs" in order to be deployed (although continued and expanded investment in scientific research can contribute further solutions not yet considered above). The challenge today is to drive the scale-up, maturation, cost reduction, and steady improvement of the full suite of low-carbon solutions. The history of successful energy innovations points the way forward.

Over the past decades, the United States has seen precipitous declines in the cost of five key technologies: wind power, solar power, shale gas, light emitting diodes (LEDs), and lithium-ion batteries for electric vehicles and grid-connected electricity storage (Trembath, 2012; DOE, 2015a). Deployment of these technologies has helped to bring about the bulk of emissions reductions to date and has transformed the economics of decarbonization. In each case, these remarkable trends were influenced by similar processes involving both proactive public investment in research, development, and demonstration (RD&D) *and* the creation of markets to hasten early adoption and ignite private sector innovation and competition through incentives and standards. Examples include the unconventional gas tax credit for shale gas, production and investment tax credits for wind and solar, utility rebate programs for LEDs, and fuel economy and zero emissions vehicle standards and electric vehicle subsidies for lithium-ion batteries. Thanks to prior decades of investment and policy, all five of these technologies went from expensive "alternative energy" to cost-competitive, mainstream energy choices that are transforming the electricity, buildings and appliances, and transportation sectors and will enable cost-effective and sustained reductions in GHG emissions over decades to come. Now, even as the United States targets deployment of these technologies at scale, the task remains to use this same successful engine of innovation to complete the net-zero carbon toolkit.

2. **Changes in energy expenditures during a net-zero transition are manageable, and less than historical expenditures.**

Under a business-as-usual scenario, U.S. energy consumers across residential, commercial, industrial, and other sectors are likely to spend more than $1 trillion annually on energy services between now and 2050 (EIA, 2019). This level of spending, including investment dollars that underpin it, provides an opportunity to leverage and redirect investment and expenditures toward a clean energy system.

Historical expenditures on energy ranged from 5.5 percent to nearly 14 percent of gross domestic product (GDP) for much of the period from 1970 through 2018 (EIA, 2020d).

Global and domestic spikes in the price of natural gas and oil have historically driven energy expenditures to the higher end of the range (as high as 9.6 percent of GDP as recently as 2008 [EIA, 2020d]). These spikes have exposed U.S. consumers and the economy to risks that could be substantially insulated if the nation were to build a net-zero emissions economy.

Multiple studies estimate that net-zero emissions could be achieved while spending roughly 4–6 percent of GDP on energy in total (Haley et al., 2019; Larson et al., 2020). Energy system expenditures in a net-zero emissions economy are likely to be higher than a business-as-usual pathway—Princeton's Net-Zero America study (Larson et al., 2020)

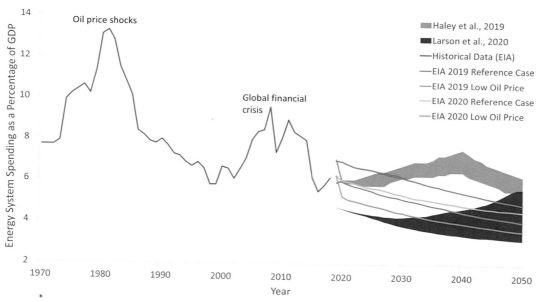

FIGURE 2.3 Historical energy system costs as a percentage of gross domestic product (GDP), with representations of the ranges of projected energy system costs under two different net-zero studies (Haley et al., 2019; upper, orange, and Net-Zero America (Larson et al., 2020; lower, purple) as well as four Energy Information Administration (EIA) projections: Annual Energy Outlook 2019 Reference (green) and Low Oil and Gas Price (light blue) cases and Annual Energy Outlook 2020 Reference (gray) and Low Oil Price (yellow) cases. These EIA projections illustrate the wide variation in energy system spending as proportion of GDP owing to unpredictable fluctuations in the prices of oil and gas, and explain the variation between the two studies (in modeling reference scenarios, Haley et al. used the AEO2019 reference oil and gas price scenario and Larson et al. used the AEO2019 low oil and gas price scenario). The ranges from the two studies are bound by each study's highest and lowest cost cases. In Haley et al. (2019), the high cost bound is in the low land negative emissions technologies (NETs) case (a scenario with a lower uptake of carbon in land sinks, resulting in a more restricted energy system-wide emissions budget), and the low cost bound is in the low electrification case through 2040 and the no new nuclear case 2040–2050. In Net-Zero America (Larson et al., 2020), the high cost bound is in the high electrification, 100 percent renewables case, and the low cost bound is in the high electrification and high-electrification, low renewables cases.
SOURCE: Data from EIA historical data; EIA (2019); EIA (2020); Larson et al. (2020); and Haley et al. (2019).

estimates cumulative incremental cost (net present value, NPV) of $4 trillion to $6 trillion from 2020 to 2050 relative to a reference case. However, adopting a net-zero economy in the United States would reduce the risks of spikes in fossil markets and reduce the share of economic activity spent on energy services relative to today's levels, while also eliminating the U.S.' ongoing contributions to climate change. Estimates of the incremental cost of a net-zero transition have been decreasing over time as the costs of clean energy technologies (e.g., wind, solar, and electric vehicles) have been declining, indicating that innovation can further decrease the costs of the clean transition.

3. **A net-zero economy requires fundamental shifts in our energy systems. The success of any pathway requires high levels of public acceptance and is bounded by societal constraints and expectations.**

Any pathway to decarbonization entails fundamental shifts in the way Americans power their homes and economies, produce goods, deliver services, transport people and goods, and manage public and private lands. This transition is bounded by societal expectations of reliability and costs of energy services and products, considerations of energy access and equity, uncertainties in the pace of technology development and deployment, and regulatory and market barriers to new technologies (EFI, 2019). The energy system has considerable inertia, aversion to risk, and market, finance, and regulatory structures that favor incumbents. Previous experiences have demonstrated that widespread adoption of new technologies is facilitated by perceived value, clear communication, and consumer incentives. For example, much of the success of the ENERGY STAR program can be attributed to its recognizable and easily understandable labeling and purchase incentives, in addition to consumer desire for improved energy efficiency (EPA, 2017). Similarly, Tesla offers vehicles that have both desirable performance features as well as decarbonization benefits. Societal preferences and policy, regulatory, and investment environments will constrain and shape the transition (EFI, 2019). These ideas are further discussed in Chapter 3.

4. **Long lifetimes and slow stock turnover of energy infrastructure and equipment limit the pace of the transition.**

Slow stock turnover in buildings, industrial facilities, and other long-lived assets leaves little room for delay and few opportunities to replace or repurpose existing infrastructure for a low-carbon energy system (Figure 2.4). Deep decarbonization can be achieved without retiring existing equipment and infrastructure before the end of their economic lifetime, which reduces the cost of the transition (Williams et al., 2014). However, long-lived infrastructure, such as power plants, buildings, and many industrial facilities and equipment, has only one natural replacement cycle before midcentury. As these assets are replaced, the new equipment must be consistent with

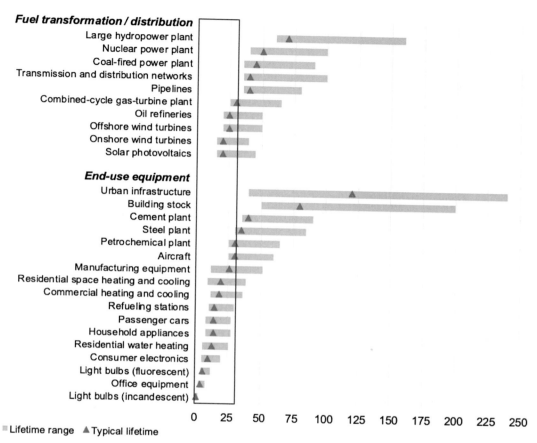

FIGURE 2.4 Typical lifetimes for key energy sector assets. The operating lifetime of some energy assets can exceed several decades, slowing the pace at which they can be replaced with cleaner and more efficient technologies. As shown by the box representing the 30-year period remaining until 2050, many assets, such as power plants, pipelines, building stock, industrial plants and equipment, and even aircraft and HVAC systems will have few natural opportunities for clean replacement before 2050. SOURCE: Adapted from IEA (2020a). Energy Technology Perspectives 2020. All rights reserved; as modified by the National Academy of Sciences.

the net-zero transition path in order to achieve net zero at the lowest total cost. Failure to replace retiring infrastructure with efficient, low-carbon successors will either result in the inability to meet emission-reduction targets or require early retirement of the replacement equipment, leading to sunk costs and stranded assets.

Recent studies see the 2020s as the time to build out enabling infrastructure for the net-zero transition and end most new investments in infrastructure to transport fossil fuels (e.g., pipelines) (Williams et al., 2018; Farbes et al., 2020; Kwok et al., 2020). Required infrastructure developments include electric vehicle (EV)-charging infrastructure for vehicles and long-distance high-voltage transmission lines (Podesta et al., 2019; Haley et al., 2019; Phadke et al., 2020), as well as hydrogen transport and storage infrastructure and with the need to plan for CO_2 infrastructure, including pipelines and storage, to come online during the 2030–2035 period (Larson et al., 2020). Efforts to site and permit new infrastructure projects must be initiated soon, given the challenges associated with obtaining permits and the long build-times. For example, new transmission lines, which are needed to connect renewable resources to areas of high electricity demand, can take as long as 16 years, and an average of 8–10 years, to site and permit (Reed et al., 2020).

5. **Repurposing existing fossil fuel infrastructure can reduce the overall costs of the transition while reducing the potential for stranded assets and workers.**

Repurposing existing energy infrastructure could play a key role in enabling a clean energy future by reducing the overall costs of the transition to clean energy, as well as mitigating likely opposition to the needed transition by reducing the potential for stranded investments and workers (EFI, 2019). For example, upgrading or converting natural gas pipelines to carry hydrogen/natural gas blends or 100 percent hydrogen could help retain the use of those pipelines in a low-carbon energy system, avoiding the need for more costly and difficult-to-site new builds while also preventing stranded assets for pipeline owners and preserving jobs in natural gas transmission and distribution utilities. Using residual oil and gas basins for permanent underground storage of carbon dioxide could help oil companies transition into carbon management utilities. Maximizing the use of existing infrastructure would help create greater buy-in for companies and their employees who benefit from the current carbon-intensive economy.

6. **A net-zero economy is very different from one with more modest reductions. Near-term actions can avoid locking in suboptimal resources.**

Reaching net-zero emissions is much more challenging and requires a different set of low-carbon resources than a system with more modest reductions. For example, modest emissions reductions in the power sector (e.g., 50–70 percent CO_2 reductions) can be achieved with deployment of natural gas-fired power plants. However, transitioning to near-zero emissions from electricity generation requires replacing the vast majority of fossil fuel power plants or equipping them with carbon capture technologies (Jenkins et al., 2018). Similarly, moderate transportation sector reductions can be achieved by blending conventional biofuels with petroleum-based transportation fuels. However, there is strong agreement in the literature that decarbonizing transportation entails the phase-out of internal combustion engine (ICE) vehicles and replacement with electric drivetrains, including battery electric and fuel cell vehicles. Policies that produce incremental reductions in emissions without facilitating transformation can lead to technology lock-in and emissions cul-de-sacs that make deep decarbonization by midcentury unattainable (Williams et al., 2014).

7. **Different decarbonization scenarios reflect different societal preferences regarding the mix of policies and technologies they employ. These scenarios can be assessed by technology mix, cost, resource needs, infrastructure buildout rates, stranded investments, jobs created and lost, societal impacts, and a suite of other factors.**

Decarbonization pathways differ in their varied mixes of policies and the central technologies upon which they depend. Some pathways are constructed using least-cost models that deploy or retire energy infrastructure based on the lowest cost of meeting energy demand without emissions. Least-cost models generally employ a broad range of zero-carbon technologies, although such models may not account for permitting and siting, regulatory, financing, or other barriers. Sometimes, lowest-cost pathways from a techno-economic or engineering perspective overlook costly impacts on certain communities or minimize or ignore friction in markets that make it difficult to accomplish those least-cost approaches.

Some other pathways are constructed using a preferred set of technologies, such as 100 percent renewables scenarios (in the electricity sector or economy wide). Still others are developed by envisioning different policy pathways, such as pathways that rely heavily on technology-neutral carbon prices or clean energy standards.

In general, decarbonization modeling finds that scenarios that constrain available technology options result in higher overall mitigation costs than scenarios that are technology neutral. For example, the Intergovernmental Panel on Climate Change (IPCC) Fifth Assessment Report determined that mitigation costs increased by 138 percent in models with no CCS (IPCC, 2014), and Sepulveda et al. (2018) found that decarbonizing the electricity sector is 11 to 163 percent more expensive if all clean-firm power generation technologies such as nuclear, CCS, and bioenergy are excluded.

More difficult to quantify, but just as important, scenarios that remove viable options generally present a greater risk of failure, as they depend more heavily on scale-up of favored technologies without impediment by any social, financial, regulatory, or other barriers. An effective risk-management strategy would hedge against likely failure modes by investing in low-carbon technologies or strategies that may prove unnecessary in a favored scenario, but provide critical alternatives should one or more bottlenecks slow progress.

8. **There are many pathways to zero emissions, and they share several core features.**

All plausible pathways to zero emissions share core features: decarbonizing electricity; switching to electricity and other low-carbon fuels for energy services in the transportation, industry, and buildings sectors; increasing energy efficiency in each of those sectors, in the power sector, and in materials; increasing carbon sequestration, and reducing emissions of non-carbon climate pollutants.

In particular, there is strong agreement among deep decarbonization studies on the following points:

- *Energy and materials efficiency:* One of the lowest-cost decarbonization opportunities helps to reduce the overall need for low-carbon fuels and electricity, and will continue to be important across all economic sectors through the next 30 years (Williams et al., 2014; White House, 2016).
- *Zero-carbon electricity:* The electric power sector should cut emissions faster and deeper than other sectors of the economy in order to meet economy-wide targets, owing to the comparative ease and wide range of zero-carbon generation options (Kriegler et al., 2014; White House, 2016; Morrison et al., 2015; Williams et al., 2014; Krey et al., 2014).
- *Electrification and fuel switching:* Electrification of energy services where possible—for example, space and water heating in buildings, light-duty cars and trucks, and some industrial processes—is key to further reducing the use of fossil energy in the end-use sectors (Kriegler et al., 2014; White House, 2016; Morrison et al., 2015; Williams et al., 2014; Jacobson et al., 2015; Steinberg et al., 2017). Zero- and low-carbon fuels can then meet much of the remaining demand for liquid and gaseous fuels (de Pee et al., 2018; ETC, 2018; Davis et al., 2018).
- *CCS:* Important for mitigating industrial process emissions, CSS may also be a useful option for the power sector (IPCC, 2018; de Pee et al., 2018; Rissman et al., 2020; Friedmann et al., 2019; ETC, 2018; Sepulveda et al., 2018).
- *Non-CO_2 gases:* These are more challenging to address, although options exist to transition away from hydrofluorocarbons (HFCs) in refrigeration and cooling, and to minimize emissions of methane and nitrous oxide (IPCC, 2018).
- *Negative emissions:* Enhancing carbon sequestration through land sinks and negative emissions technologies is important to counter residual emissions from non-CO_2 gases and hard-to-abate energy sectors that are impossible or prohibitively expensive to eliminate completely (IPCC, 2018; NASEM, 2019).

9. **Decarbonization studies converge on similar near-term (2021–2030) specific actions needed to put the United States on a path to net-zero emissions by 2050.**

Feasible decarbonization pathways are very similar in the first 10 years and diverge only in later years. This first report focuses on near-term priority decarbonization actions for 2021–2030 that are robust across many scenarios or retain optionality in the face of uncertainty about the final decarbonization pathway. The committee's assessment of decarbonization approaches and pathways from 2021–2050 will be discussed in its second report.

Analyses that model pathways to net-zero emissions in 2050 agree that in the next 10 years, the United States must:

- *Improve efficiency of material and energy use* by 15 to 19 percent in the industrial sector (Ungar and Nadel, 2019; Larson et al., 2020); 20 to 30 percent in the building sector (Ungar and Nadel, 2019; Mahajan, 2019a); and 10 to 15 percent in the transportation sector between 2021 and 2030 (Larson et al., 2020).
- *Electrify energy services that directly use fossil fuels* at the rate of 10 to 50 percent of new light-duty vehicles, and heat pump electrification of space heating and water heating in 15 to 25 percent of residences, with all new construction to be fully electric in order to achieve >50 percent of building energy supplied by electricity by 2030 (up from ~44 percent today). Industrial boilers fueled with natural gas are replaced with electric as they retire (Ungar and Nadel, 2019; Mahajan, 2019a,b; Rissman, 2019; Larson et al., 2020).
- *Increase clean electricity generation* from 37 percent of U.S. electricity in 2020 to roughly 75 percent by 2030 through expanding generation capacity of wind (~250–300 GW) and grid-scale solar (~280–360 GW) (Larson et al., 2020; Phadke et al., 2020). Coal retirements continue or accelerate and contribute ~1 $GtCO_2$ emissions reduction by 2030. These analyses assume that most existing nuclear capacity should be preserved (and/or expanded with upgrades), with studies ranging from 11 GW retirement to 5 GW addition in 2030 (Haley et al., 2019; Larson et al., 2020).
- *Build no new long-lived fossil fuel infrastructure* (such as pipelines) that cannot be repurposed for use in a net-zero economy, and instead build network infrastructure to enable net-zero energy transition. This assumes that the nation must begin the siting, permitting, and building of high-voltage transmission lines (up to ~60 percent increase in total GW-miles of capacity [Larson et al., 2020]), electric-vehicle charging infrastructure (Haley et al., 2019; Podesta et al., 2019), with ~1–3 million Level 2 chargers and ~100,000 DC faster chargers

(Larson et al., 2020), and the planning for siting, permitting, and construction of hydrogen storage and transport networks and trunk pipelines for a national CO_2 transport system (as much as 12,000 miles by 2030 [Larson et al., 2020]).
- *Continue to demonstrate and improve CCS and capture as much as 65 MMT CO_2 per year at industrial and power facilities, equivalent to about 5 large cement facilities, 5–10 methane reforming hydrogen production facilities, and 5–10 gas power plants with CCS* (Larson et al., 2020). Begin demonstration of direct air capture (DAC), and build out DAC capacity of 9 MMT CO_2 per year by 2030 (Larsen et al., 2019).
- *Invest in RD&D and create niche markets via incentives and standards* to drive innovation, maturation, and improvement of a range of nascent technologies including for hydrogen production from biomass gasification, direct air capture, low-carbon or carbon-sequestering materials, low-carbon synthetic fuels, advanced nuclear, and other low-carbon energy technologies (Haley et al., 2019; Larson et al., 2020; NASEM, 2019; Podesta et al., 2019).

This report builds on the existing, robust literature on possible pathways to deep decarbonization. Metrics for three of the most recent and comprehensive studies are reported in more detail in Table 2.1. The scenarios analyzed in these studies projected energy demand, share of non-emitting electricity, share of electricity in final energy demand, energy productivity, and scale of CCS, land sinks, hydrogen production, impact of non-CO_2 gases, building energy intensity, and EV share. Table 2.1 compares both their 2030 and 2050 results. Although these studies and models rely on different assumptions, data, and methods, the comparison in Table 2.1 illustrates their coherence in the first 10 years in particular.

> **10. New open-source energy system optimization models need to be developed to further study transitions, trade-offs, and opportunities in net-zero energy systems.**

No model currently exists in the public domain that is capable of modeling all major elements of a net-zero system at the requisite level of detail to analyze: deep reductions in energy demand through efficiency in vehicles, appliances, buildings; flexible central-station and distributed resources (including flexible demand) at dispatch time scales; power flows and realistic expansion of local and high-voltage electricity networks; gas and liquid fuels production, transportation, storage, and consumption; CO_2 capture, pipelines, use, and sequestration; and non-CO_2 greenhouse gases and carbon sinks. The primary technical impediment to developing such a model is computational constraints, because the model must simultaneously optimize decisions across all sectors, at high temporal resolution (to capture flexibility needs and

TABLE 2.1 Relevant Metrics/Indicators Across Three Separate Decarbonization Studies

Key Metric	2015[a]	2030			2050		
		Energy Innovation[b]	Deep Decarbonization Pathways Project[c]	Net-Zero America Project[d]	Energy Innovation	Deep Decarbonization Pathways Project	Net-Zero America Project
Final Annual Energy Demand (quads)	97	129	80	64–67	125	65	50–56
Percent Non-emitting electricity	18	60	55	62–77	100	85	98–100
Electricity share of final energy demand (percent)	28	44	32	21–25	73	60	38–51
Energy productivity of GDP ($ economic output per energy) ($ billion/quad)[e]	185.5	182	293	350–367	272	524	609–682
Carbon capture (MMT CO_2/yr)	0	30.6	ND	65–197	26	775	690–1760
Land sinks (MMT CO_2/yr)	760	245	1050	750	630	1050	850
Hydrogen production (quads/yr)	0.74[f]	2.5	<1	0.95–1.9	5.5	<1	7–18
Non-CO_2 gases (MMT CO_2e/yr)	1264	1243	ND	1090	587	ND	1020

continued

TABLE 2.1 Continued

Key Metric	2015[a]	2030			2050		
		Energy Innovation[b]	Deep Decarbonization Pathways Project[c]	Net-Zero America Project[d]	Energy Innovation	Deep Decarbonization Pathways Project	Net-Zero America Project
Building energy demand (quads/yr)	18	17	16.4	18–19	11	13	13–15
EV share of light-duty vehicle stock (percent)	1	47	44	6–17	100	100	61–96

[a] EIA 2019a, 2020a; EPA, 2019, 2020; White House, 2016.
[b] Energy Innovations, 2020.
[c] Haley et al., 2019. Follows EIA projections for economic growth and increased consumption of "energy services." Assumes rapid adoption of electrification technologies and high-efficiency technologies where the end-use is already electric (i.e., refrigeration) or where complete electrification is infeasible. Adoption rates of these technologies accelerate through 2030, with the stock of these technologies lagging but making steady progress through 2050. Assumes an enhanced land sink 50 percent larger than the current annual sink. Assumes that nuclear plants already in operation will be operated and retired based on the schedule in the 2017 *Annual Energy Outlook*.
[d] Larson et al., 2020.
[e] Calculated using data from PWC, 2017.
[f] D. Brown, 2016. U.S. Hydrogen Production—2015. CryoGas International.

NOTE: These studies, while conducted with different modeling frameworks and assumptions, find commonality in the near term and greater divergence in the long term. This comparison illustrates agreement in the literature regarding near-term actions to begin a long-term energy transition and underscores the importance of actions that maintain or enhance optionality in the long term. ND = not determined.

impacts of variable renewable electricity production), and with sufficient geospatial detail to capture complex variations in demand, siting limitations, and local policies and to provide actionable insights to inform real-world decision making. New tools and ways of thinking about energy system models will be required to overcome these barriers. The United States should invest in the development of an ecosystem of open-access modeling tools and open-source data to accurately parameterize these models to help plan the transition to net zero and to better represent the universe of possible net-zero transitions.

THE FIRST 10 YEARS: FIVE CRITICAL ACTIONS

This report identifies 10-year actions that are robust across decarbonization pathways. The committee emphasizes strategies that are (1) "no-regrets" actions that would be needed regardless of the final path taken or (2) that retain "optionality" and flexibility so that the United States can take advantage of technological advances, mitigate risks that could derail primary strategies, and avoid stranded actions. Such an approach is also important in light of uncertainties in technology, support for climate policy, differences in regional energy resources or stakeholder preferences, and future climate impacts. The final report focusing on a longer time period will need to consider more strongly methods for planning and policy making under deep uncertainty (Marchau et al., 2019; Mathy et al., 2016; Waisman et al., 2019; Bataille et al., 2016). However, identifying a strategy for 2021–2030 is easier than it sounds, because feasible paths for near-term emissions reductions and early investment in long-term potential strategies are very similar in the first 10 years and diverge only in later years.

For these strategies, the committee has provided estimates of the pace and depth of needed technology deployment and action, in order to provide the order of magnitude of changes warranted in a no-regrets strategy and set of actions in the next 10 years. A selection of these is summarized in Table 2.3 below.

1. Invest in energy efficiency and productivity.

Energy and materials efficiency is one of the most cost-effective near-term approaches to reduce energy demand and associated emissions. This approach includes adopting developing technologies and processes that increase fuel efficiency of vehicles (on-road and off-road, including farming equipment); increasing the efficiency of building enclosures as well as installing efficient appliances and equipment in buildings; enhancing energy productivity in manufacturing and other industrial processes and in the power generation fleet; and improving systems efficiencies from greater energy system integration. Demand efficiency and materials efficiency measures (e.g., recycling and reuse) are also included in this category. Priority actions in the 2021–2030 time frame include the following:

- **Buildings:** Reduce building space conditioning and plug load energy use by 3 percent per year for existing buildings from a 2018 baseline, to achieve a 30 percent reduction by 2030. Meet the Architecture 2030 goal of carbon neutrality for all new buildings, developments, and major renovations by 2030 (Architecture 2030, n.d.). These targets may be met by implementing a combination of sustainable design strategies, generation of on-site renewable energy, and/or purchasing (20 percent maximum) of off-site renewable energy. It is also critical to work toward maximum conditioning goals for new construction that reflect passive house site energy standards of 5–60 kBtu/ft^2/year (depending on climate and building type), with plug loads held to 3000–4000 kWh/year per household, and peak demand capped under 10 W/m^2 (3.2 Btu/ft^2). As addressed in Wright and Klingenberg (2018), it is essential to reduce peak loads in addition to operational demands through conservation and load shifting in both new and existing buildings. Incorporate district heating, where feasible.
- **Transportation:** Increase energy productivity by encouraging shifts in transportation from single-occupancy light-duty vehicles (LDVs) to multi-occupancy vehicles, public transit, cycling, and walking (although historically, these shifts can be difficult or costly to achieve). Shift on-road trucking to freight rail. Steadily improve the fuel efficiency of new ICE vehicles—especially important for the medium-duty vehicle/heavy-duty vehicle (MDV/HDV) sectors, as well as planes, ships, and trains, which are more difficult and/or expensive to power with electricity. Encourage flexible and remote work patterns. Invest in improved real-time traffic control, introduce automated vehicles for smoother traffic flow and less congestion from crashes, and reduce travel through telework and mixed-use development. Efficiency improvements could reduce emissions by 10 to 30 percent over the next few decades (Lah, 2017). Between 2007 and 2017, average annual improvement in LDV fuel economy was 1.9 percent per year, and this could be continued in the next decade and extended to trucks with appropriate policies in place (Table 4.1 of Davis and Boundy, 2020). The aviation and maritime industries have also established goals and policies for substantial GHG reductions, primarily through efficiency improvements in the use of alternative fuels (USG, 2015; ICCT, 2018).
- **Industry:** Deliver 25 percent of the potential industrial sector energy efficiency reductions (3 quads, 117 million tons CO_2 reduction) by 2030 (Ungar and Nadel, 2019). Achieve 3 percent per year sustained improvement in industrial energy

productivity (i.e., dollar of economic output per energy consumed) and improving materials efficiency by minimizing/recycling waste by 10 percent, and advancing waste heat recovery/reuse to improve energy efficiency of process equipment such as furnaces by 10 percent. Optimize systems and promote energy and materials management—for example, strategic energy management (SEM)—across all industries and all size companies, advance smart manufacturing, and institute circular economy strategies.
- **Embodied energy in products and building materials:** Increase materials and water efficiency to reduce associated energy and GHG inputs. Decrease high-carbon-intensive building and infrastructure materials with goals to reduce carbon intensity by a minimum of 30 percent and to pursue carbon-sequestering alternatives.

2. **Electrify energy services in the buildings, transportation, and industry sectors.**

Electrification of energy services, in tandem with decarbonization of electricity generation, has emerged as a core element in nearly all deep decarbonization scenarios. The greatest near-term (2021–2030) potential for electrification is in the buildings and transportation sectors. In buildings, electric heat pumps for space conditioning and water heating can help lower carbon emissions compared to fossil systems. Among LDVs, electric vehicles are projected to reach cost-parity with internal combustion engine vehicles in the next decade and, in conjunction with relatively low-carbon electricity, will also reduce emissions. Some potential exists for electrification of industrial processes, although electrification technologies for the industrial sector are at a relatively early stage of development and play a greater role beyond the 2030 time frame, as electrification technologies mature, decline in cost, and are demonstrated at scale.

- **Buildings:**
  - *Space heating:* Deploy high-efficiency heat pumps in ~25 percent of current residences by 2030 (25–30 million households) and 15 percent of commercial buildings. Focus on stock turnover and new builds in climate zones 1–5,[1] planning for 100 percent of sales by 2030.
  - *Hot water:* Switch to heat-pump hot water heaters when existing stock reaches end of life, ramping up to 100 percent of new sales by 2030.

---

[1] Climate zones are based on heating degree days, average temperatures, and precipitation. Climate zones 1–5 cover all of the United States except for the "cold," "very cold," and "subarctic" regions that include Alaska, the northern half of Rockies, the Upper Plains states, Minnesota and Wisconsin, northern Michigan, upstate New York, and the northern half of New England (DOE, 2015b).

- **Transportation:**
  - *Electric vehicles:* Approximately 50 percent of new vehicle sales across all vehicle classes (light, medium, and heavy duty) and 15 percent of on-road fleet will be electric vehicles (with some fuel cell EVs in the MDV and HDV subsectors) by 2030. This includes approximately 50 million LDV cars and trucks and 1 million MDV and HDV trucks and buses. Invest in more electrified train services and aircraft. Ports and airport taxiing should be electrified.
  - *Renewable transportation fuels:* Expand power to liquids opportunities for post-2030 by developing regionally based pilot production facilities.
- **Industry:**
  - Develop and deploy options to decrease emissions from process heat production, including a significant proportion of electric technologies. As opportunities arise for replacement of legacy equipment, advance the use of low-temperature solutions such as heat pumps, infrared, microwave, electric and hybrid boilers, and other options as described in Rightor et al. (2020, Appendix A).
  - Deploy tens of GWs of electric boilers to supply low- and medium-temperature heat for various industrial processes whenever electricity cost, economics, and non-energy benefits can justify replacement. In some applications, electric boilers can be installed alongside existing gas boilers, enabling hybrid use of electricity to displace fossil fuels when electricity supply is abundant and costs are low.
  - Deploy 1–2 GW of advanced industrial heat pumps (IHPs), with early development/ demonstrations at industrial clusters to lower barriers, for a range of process heat, drying, evaporator trains, and other applications lowering CO_2 emissions with the electricity coming from low-carbon sources.

3. **Produce carbon-free electricity.**

The electric sector plays a critical role in decarbonization, both in terms of reducing GHG emissions from electricity production and use and for supporting the decarbonization of other sectors. Since 2005, the share of electricity from zero-carbon emitting sources—including nuclear power, hydropower, wind, solar, biomass, and geothermal—has increased from 28 percent to 37 percent. This growth comes primarily from wind and solar, as cost reductions and policy incentives have combined to drive deployment (even as other zero-carbon emitting technologies have declined or remained stagnant). Wind or solar power is now the cheapest source of new electricity generation in 34 percent of U.S. counties, based on levelized cost of electricity and considering regional differences in capital costs and fuel delivery prices (UT-Austin, 2020).

The 2020s are a key decade to build out the electric transmission and distribution infrastructure needed to accommodate flows from and access to these commercially ready new zero-carbon resources.

- **Electricity generation and storage:**
    - *Carbon-free electricity:* Roughly double the share of U.S. electricity generation from carbon-free sources from 37 percent today to roughly 75 percent nationwide by 2030.
    - *Wind and solar power:* Deploy ~250–300 GW of wind (~2–3× existing capacity) and ~300 GW of solar (~4× existing) by 2030, supplying approximately 50 percent of U.S. electricity generation (up from 10 percent today). To reach this level, the sustained annual pace of wind and solar capacity deployment must match or exceed record annual rates to date from 2021–2025 and accelerate to roughly double that rate in the 2026–2030 time frame.
    - *Coal power:* Manage continued (or accelerated) retirement of existing coal-fired power plants, including associated operational reliability and local economic transition challenges and impacts.
    - *Nuclear power:* Preserve existing nuclear power plants wherever safe to continue operation as a foundation for growing the carbon-free share of electricity generation. The deployment of small modular reactors may occur by the late 2020s and provide additional clean electricity generation.
    - *Natural gas power plants:* Modest decline in gas-fired electric generation (10 percent–30 percent) and capacity is roughly flat nationally through 2030 to maintain reliability as coal (and some nuclear) units retire, and to provide system flexibility alongside wind, solar, and storage, while avoiding new commitments to long-lived natural gas pipeline infrastructure.
    - *Energy storage:* Deploy 10–60 GW / 40–400 GWh of intraday energy storage capacity (e.g., battery energy storage) through 2030 to reduce need for infrequently utilized peaking power plants, mitigate transmission and distribution constraints, and integrate variable renewable energy. Enhanced demand flexibility (e.g., through real-time pricing, demand response programs, and aggregation and control of flexible loads such as electric vehicle charging) can directly reduce the scale of battery storage required.

4. **Plan, permit, and build critical infrastructure and repurpose existing energy infrastructure.**

In the 2020s, efforts must begin to build out enabling infrastructure for the low-carbon transition. These will include EV-charging networks (to enable vehicle electrification); long-distance high-voltage transmission lines (to bring remote power resources to population centers, because high-quality renewable sources are often not located near major load centers); upgrades to distribution grid upgrades to enable electrification of heating and transport; and renewable fuel (e.g., hydrogen) transport and storage infrastructure. Planning and siting for a national CO_2 pipeline system should begin immediately, and various developments in the first half of the decade will determine whether CO_2 infrastructure, including pipelines and storage, will need to be built at scale by 2030 or the middle of the next decade.

- **Transportation:**
  - *Charging infrastructure:* Proactive build-out of EV charging infrastructure to facilitate greater adoption of EVs, including 2–3 million Level 2 chargers and at least 100,000 DC fast chargers by 2030. This infrastructure should be a mix of private and public ownership and operation, including fleet operators.
  - *Investment* in vehicle connectivity and real-time control infrastructure.
- **Electricity transmission and distribution:**
  - *Electric transmission:* Strengthen and expand U.S. long-distance electricity transmission by identifying corridors needed to support wind and solar deployment (both through 2030 and beyond, given the long siting and build timeline for transmission), which will require policy and process reforms described in Chapter 4. Leverage opportunities to reconductor existing transmission lines at higher voltages and take advantage of existing rights of way and dynamic line rating to enhance existing transfer capacity. Increase overall transmission capacity (as measured in GW-miles) by about 40 percent by 2030. Incorporate new materials to reduce losses and increase efficiency.
  - *Electric distribution:* Strengthen distribution-system planning, investment, and operations to allow for greater use of flexible demand and distributed energy resources for system needs, improve asset utilization in the distribution network, and efficiently accommodate up to an approximately 10 percent increase in peak electricity demand from EVs, heat pumps, and other new loads during the next decade. Prepare for more-rapid electrification and peak demand growth after 2030.
  - *Expand smart grids:* Expand automation and controls across electricity distribution networks and end-use devices by increasing the fraction of electricity meters with advanced two-way communications capabilities from about half to 80 percent. Smart grid expansion will enable greater demand response of EV charging, space and water heating loads, and cooling energy storage for air conditioning buildings. It will also allow the use of a

variety of smart home and business technologies that can increase energy efficiency while reducing consumer costs. Further development of the broadband network across the country is required in order to enable these smart grid expansions. Such actions could also spur economic development and potentially reduce transportation-related carbon emissions by facilitating telework.

- **Fuels:**
  - *Expand hydrogen infrastructure*, including transmission and distribution.
  - *Leverage the current natural gas pipeline infrastructure* to operate with 5 percent hydrogen (on an energy basis), with appropriate user retrofits. Complete one or more demonstrations of large-frame combustion turbine operations consuming greater than 20 percent hydrogen (by energy content) on an annual basis through typical operational cycles for multiple years to reduce technology risk and identify longevity and operability challenges with high hydrogen/natural gas blends.
  - *Build connections from points of H_2 generation* (via electrolysis or other renewable sources) to the user base, current hydrogen delivery infrastructure, and natural gas distribution system (for blending purposes). Maximize opportunities to utilize and repurpose existing gaseous and liquid fuel transmission, distribution, and logistics infrastructure. Expand hydrogen refueling for medium- and heavy-duty vehicles. Hydrogen networks will likely be regional in scope, given the ability to cost-effectively produce hydrogen in most parts of the country from a combination of electrolysis, natural gas reforming with CCS, and biomass gasification.
- **Industry:**
  - *Define infrastructure requirements to deliver on industrial needs* (e.g., interconnections, substations, high-voltage lines, storage, and grid energy flows). Pursue these capacity improvements in collaboration with utilities and industry, again starting with clusters.
  - *Build capability, market pull, and lower costs for hydrogen use* in iron and steel, chemistry, and refining, targeting 2 percent of combined energy and fuel use by 2030 to kick-start future increases.
- **Carbon capture, utilization, and sequestration (CCUS):**
  - *CCUS network development:* Set the foundation for large-scale CCUS by planning for the location and timing of an "interstate CO_2 highway system" or trunk line network, and determine by mid-decade whether construction of trunk lines needs to be completed by 2030 or 2035 (~10,000 miles, up from 4,500 miles today). Regional clusters can be a starting point of a larger, interconnected network. This network will connect the

high CO_2 supply that needs to be abated long term (50 to 75 MMT CO_2 per year by 2030 and as much as 250 MMT CO_2 by 2035) to regions of high CO_2 use potential or storage. Development of a CO_2 network could involve repurposing existing natural gas or oil pipeline infrastructure or rights-of-way.
- *Reservoir characterization:* Characterize sustained CO_2 injection rates that can be achieved across each of the major CO_2 sequestration basins and identify by 2030 high injection rate locations suitable for injection of approximately 250 million metric tons of CO_2 per year.

5. **Expand the innovation toolkit.**

For some sources of emissions, and particularly those in harder-to-abate sectors, low-carbon alternatives are still in the pilot stage or remain nascent industries. For these sectors, near-term opportunities for emissions reduction are limited to improving energy efficiency, materials efficiency, demand management, and other tools that reduce—but cannot completely eliminate—the emissions intensity of these sectors. Maturation, improvement, and scale-up of an expanded set of carbon-free alternatives will be needed as near-term emission reduction opportunities are exhausted. Bringing new energy technologies to market can take 20–70 years from the first prototype, and driving maturation and cost declines for nascent industries proceeds over a decade or longer time scales. Therefore, proactive RD&D and market creation efforts are needed in the 2020s to develop, improve, and scale up nascent low-carbon energy technologies, including the following:

- **Electricity generation:**
    - RD&D and early market deployment for clean-firm electricity resources (e.g., advanced nuclear, CCS, enhanced geothermal, and hydrogen combustion turbines or fuel cells).
- **Industry:**
    - Develop transformative processes for utilizing low-carbon energy carriers (e.g., hydrogen) in the generation of low-carbon precursors and products (ammonia, methanol, ethylene, etc.) and as solutions for reductants (e.g., steel).
    - Develop and pursue low-carbon process heat solutions across all temperature ranges, especially providing options for mid and high temperatures.
    - Advance electrolyzer efficiency and longevity, thereby enabling lower costs and broader application of water electrolysis for H_2 and other electrolytic processes
    - Substantially increase the efficiency of separations to cut energy costs (upward of 50 percent energy spend for some processes) and introduce low-carbon separation (e.g., membranes driven by electricity).

- **Energy storage:**
    - RD&D for batteries and other energy storage technologies.
    - Improve battery storage for vehicle applications to achieve cost below $50/kWh, performance above 500 Wh/kg, a 10-year life, and several thousand cycles.
    - Improve long-duration energy storage for deployment with the electric grid and renewable energy to operate at an ultra-low cost per kWh (~ $1/kWh) and long asset life (e.g., 10–30 years).
- **Fuels:**
    - RD&D and early market deployment to reduce costs of net- zero carbon fuels, including drop-in and non-drop in fuels, to be cost-competitive with electrification. Specific areas of interest include hydrogen production from electrolysis, biomass gasification, and methane reforming with CCS, particularly early commercial deployment to drive experience and reduce costs; synthesis of hydrocarbon fuels from cellulosic biomass and H_2 and CO_2 via Fischer-Tropsch or methanation processes (e.g., "drop-in" fuels); and high-yield bioenergy crops.
- **Carbon capture, utilization, and sequestration:**
    - Develop CCUS technologies (including with support of enabling policies) for a variety of applications across the industry and power generation sectors.
    - Perform advanced characterization of geologic formations that have received little attention but may have significant impact (e.g., basalt, ultramafics,[2] and saline aquifers). Survey and analyze natural and industrial alkaline sources that could serve as a feedstock for CO_2 mineralization.
    - Integrate CCS with process heat to lower costs.
    - Continue developing and deploying more efficient capture technologies (e.g., Jacoby, 2020) and other negative emission technologies.
    - Advance direct CO_2 utilization (e.g., syngas, Fischer Tropsch, etc., with renewable H_2 and recycled CO_2).
- **Innovation to reduce infrastructure siting challenges:**
    - Increase investment in research, technology, and process/procedural solutions that reduce siting challenges with network infrastructure, including repurposing existing natural gas or oil pipelines for hydrogen or CO_2 transport, developing low-cost underground transmission lines on existing rights of way, and increasing utilization and transfer capacities of existing electricity transmission.
    - Coordination of these activities to account for the timing of demand changes for CO_2, natural gas, and oil as well as the higher pressure operation of CO_2 pipelines will be required.

---

[2] Ultramafic rock is igneous in nature.

BOX 2.1
METHODS TO LIMIT NON-CO_2 GHG EMISSIONS

The committee focused this interim report on CO_2 emissions, as directed by its task statement, but also recognizes that net-zero refers to all anthropogenic greenhouse gases (GHGs) covered by the United Nations Framework Convention on Climate Change (UNFCCC), including methane, N_2O, and fluorinated gases. The sources of non-CO_2 gases are generally more challenging to address than CO_2, in part because they are more diffuse and because some are associated with agricultural activities that cannot be fully abated. However, some reductions can be achieved through higher efficiency processes (precision agriculture to reduce N_2O, improved methane leak detection and mitigation to reduce CH_4 from fossil energy systems, etc.), and by replacing hydrofluorocarbons (HFCs) in refrigeration and air conditioning with other coolants such as CO_2. As detailed in the recent National Academies of Sciences, Engineering, and Medicine report on negative emissions technologies, the existing land sink and other low-cost agricultural and forestry options can offset any residual non-CO_2 emissions (NASEM, 2019).

Non-CO_2 emissions in the United States totaled 1,250 million metrics tons of CO_2-equivalent (MtCO_2e) in 2018 (EPA, 2020). These non-CO_2 GHGs, including methane, nitrous oxide, and fluorinated GHGs, are more effective than CO_2 at trapping heat within the atmosphere and in some cases can remain in the atmosphere for longer periods of time. Given the significant warming effect of non-CO_2 GHGs, achieving the nation's climate goals requires deep reductions in their emissions in addition to deep decarbonization strategies. In line with the IPCC Special Report on 1.5°C, this report assumes that methane emissions can be reduced by 65 percent below 2010 levels by 2050, and nitrous oxide can be reduced 25 percent (IPCC, 2018). Per the Kigali Amendment, HFCs will be reduced by 85 percent by 2045 (United Nations, 2016). With these conditions in place and utilizing various abatement strategies, total non-CO_2 emissions would decline from 1,250 MtCO_2e in 2018 to 600–700 MtCO_2e by 2050. To offset these residual non-CO_2 emissions and achieve net-zero total GHG emissions, implementation of negative emission technologies that sequester CO_2 are also required.

Non-CO_2 GHGs originate from a wide variety of sources. The main sources of methane include enteric fermentation and manure management associated with domestic livestock, natural gas systems, decomposition of wastes in landfills, and coal mining (White House, 2016; EPA, 2020). Nitrous oxide emissions are associated with agricultural soil management, stationary fuel combustion, manure management, and mobile sources of fuel combustion (EPA, 2020). The vast majority of fluorinated gases emitted are HFCs primarily used for refrigeration and air conditioning.

The energy system has the largest potential for non-CO_2 GHG mitigation, followed by the industrial, waste, and agricultural sectors (EPA, 2019). Natural gas and coal activities represent the largest contributors to non-CO_2 emissions. In natural gas and oil systems, significant mitigation of non-CO_2 emissions can be achieved through changes in operational practices, including directed inspection and maintenance. In coal mining, reduction of ventilation air methane and degasification for power generation and pipeline injection represent most of the abatement potential.

Mitigation potential from the industrial processes sector lies primarily in refrigerants, air conditioning, and N_2O abatement measures in fertilizer production. Significant mitigation potential

BOX 2.1 Continued

also exists in electronics manufacturing and aluminum and magnesium production. In the waste sector, abatement measures in landfills—including collection and flaring, landfill gas utilization systems, and waste diversion practices—and improvements to wastewater infrastructure can provide significant reductions in non-CO_2 GHG emissions. Measures applied to livestock, croplands, and rice cultivation, such as use of anti-methanogens and reduction of fertilization, provide the highest mitigation potential in the agricultural sector. Additional mitigation measures in the agricultural sector include livestock dietary manipulations like the use of propionate precursors; manure management with large-scale complete-mix digesters, covered lagoons, and fixed film digesters, and cropland strategies such as no-till practices and nitrification inhibitors.

Concurrent with the abatement measures above, implementation of negative emission technologies (NETs) and strategies is necessary to achieve net-zero emissions by 2050. A 2019 National Academies committee estimated the low-cost removal potential of NETs (less than \$20/tCO_2) in the United States at 520 $MtCO_2$, assuming full adoption of agricultural soil conservation practices and forestry management practices (NASEM, 2019). Including bioenergy with carbon capture and sequestration plants and waste biomass capture could remove another 500 $MtCO_2$ at less than \$100/tCO_2. These low-cost options of agricultural soil conservation and forestry management practices can be implemented now, and, together with the ongoing managed forest carbon sink in the United States (700 $MtCO_2$/y but declining), are enough to offset the residual non-CO_2 GHG emissions. In the long term, these practices could be supplemented by other technologies for removing CO_2 from the atmosphere (e.g., direct air capture and carbon mineralization), whose research needs were laid out in the 2019 National Academies report.

The forestry and agricultural policies necessary to create and maintain the required ecosystem sinks are not part of this interim report but will be part of the final report. However, the economy-wide price on carbon proposed in Chapter 4 should be enough through 2030 and beyond given the costs reported for land-based NETs in NASEM (2019). The required forestry effort was included as part of the Obama administration's Deep Decarbonization Report (White House, 2016) and is widely understood. The necessary policies for agricultural soils are well developed for some crop and soil combinations, but a monitoring and verification effort involving direct measurements of a statistical sample would need to be developed by the U.S. Department of Agriculture. Additionally, the National Academies committee (2019) called for an experimental effort to extend the ability to restore lost carbon in agricultural soils to all croplands and grazing lands. Private companies who seek co-benefits and carbon credits from private markets have now begun that work at the required scale. Extending the improvements for forestry management to include urban forests would not greatly add to the carbon removal potential. However, the co-benefits from increasing urban forestry are large and include reducing urban heat island effect and improving ambient air quality.

IMPACT ON U.S. ENERGY EXPENDITURES IN THE 2020s

Many recent studies estimate that from a technical point of view, the United States could transition to net zero by 2050 using only commercial and near-commercial technologies and spending a smaller fraction of the nation's GDP on energy system expenditures[3] than the country has in the past, including the past decade (see Figure 2.3). However, energy system expenditures during a net-zero transition would be significantly greater than business as usual. If technological options improve faster than considered in recent modeling studies, then the cost of decarbonization could prove lower.

Studies reviewed by the committee in this chapter (Larson et al., 2020; SDSN, 2020) indicate that cumulative incremental energy system expenditures during a net-zero energy transition would be approximately $100 billion to $300 billion through 2030, and $4 trillion to $6 trillion through 2050 beyond the $22.4 trillion in a business-as-usual baseline. (These estimates are reported on a NPV basis of cumulative total expenditures with a 2 percent real social discount rate.[4] With a 5 percent social discount rate, the impact would be $210 billion to $270 billion through 2030 and $2 trillion to $3 trillion through 2050. These estimates do not provide a commensurate indicator of the benefits of these investments.) It is important to note that these estimates of energy costs do not capture general equilibrium effects, such as changes in global oil prices. Note, however, that a net-zero transition would greatly reduce U.S. oil demand and put substantial downward pressure on prices. Nor do these cost estimates include impacts of changes in the U.S. balance-of-trade and other effects, which include both positive and negative factors.

The costs for deep decarbonization also must be considered in the context of the considerable benefits of a clean energy transition that could offset some, all, or more than the cost of the transition. There are climate benefits, new economic and employment opportunities, substantial improvements in public health, and intangible global leadership credentials. For example, Hsiang et al. (2017) estimate U.S. economic losses of 1.2 percent of GDP per 1°C temperature rise, with risk distributed unequally across the country and the poorest third of counties in the United State projected to incur the largest damages. They estimate the mitigation of economic damages of $200 billion to $300 billion annually by 2100 compared to a business-as-usual course.

---

[3] The energy system expenditures referenced here encompass both energy supply and demand, but do not include capital investments.

[4] Discount rates put a present value on future costs and benefits. Social discount rates attempt to value the cost and benefits for future generations relative to costs and benefits today.

Benefits of a net-zero transition also include reductions in premature deaths owing to reduced air pollution from fossil fuels, with the magnitude ranging by study: a reduction of 85,000 total premature deaths from air pollution over the 2020–2050 time period from decarbonizing electricity (Phadke et al., 2020); a reduction of 11,000 to 52,000 annual premature deaths from the elimination of air pollution from coal power plants (Prehoda and Pearce, 2017; Larson et al., 2020); and a reduction of up to 200,000 annual premature deaths from eliminating air pollution from fossil fuels entirely (Lelieveld and Münzel, 2019). In addition, a recent report estimated 5 million sustained jobs could be associated with electrifying most energy uses beyond an even larger initial surge of the infrastructure deployment (Griffith and Calisch, 2020), although this would be offset by the loss of about 1.6 million jobs in fossil fuel related sectors. Another recent study estimates that a net increase of roughly 1 million to 5 million jobs would be supported by energy supply-related sectors by 2050 (0.5–1 million by 2030), as total employment in wind, solar, transmission, and other growing sectors offset losses in oil, gas, and coal, in aggregate (Larson et al., 2020). The committee's task directed it to focus on mitigating emissions, and therefore these beneficial impacts are not extensively reviewed. However, it is clear there are substantial benefits of a net-zero transition.

MOBILIZING CAPITAL INVESTMENT IN THE 2020s

Figure 2.5 and Table 2.2 summarize the roughly $2 trillion in incremental capital investments that must be mobilized over the next decade for projects that come online by 2030 (i.e., total capital in service in the 2020s) to put the United States on track to net zero by 2050. This includes roughly $0.9 trillion in incremental capital investment in supply-side sectors and networks (roughly double the total capital expenditures under business-as-usual) and $1.2 trillion in incremental demand-side investments in buildings, vehicles, and industrial efficiency. It is important to note that these capital investments are not a direct cost borne by either taxpayers or energy consumers. They are investments in the U.S. economy made by both private and public sector actors. The sum of capital investments that must be mobilized in the 2020s is much larger than the increase in total consumer energy expenditures described above because capital investments are paid back through energy expenditures over many years and because investments in renewable electricity, efficient buildings and vehicles, and other capital-intensive measures offset significant annual expenditures on consumption of fuels.

Box 2.2 discusses potential synergies within the systems involved in a net-zero transition, including possible trade-offs and unintended consequences.

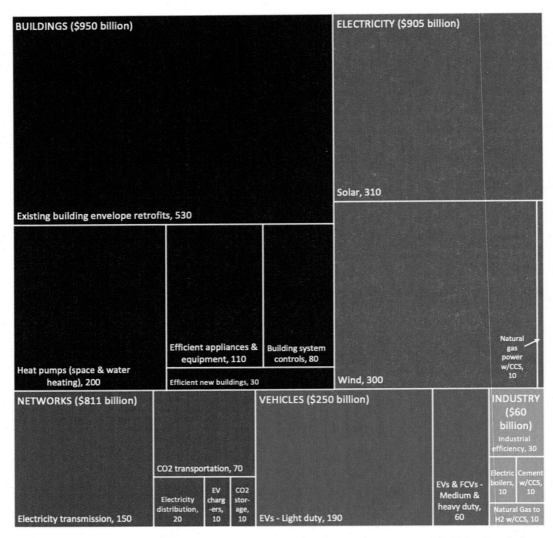

FIGURE 2.5 To put the United States on a path to net-zero emissions by 2050, roughly $2.1 trillion in incremental capital investment needs to be mobilized into the five critical actions for 2021–2030 described above. Estimates rounded to nearest $10 billion and should be treated as approximate (e.g., order of magnitude) given uncertainties. Other potentially significant changes in capital expenditures are not estimated in the above figure, including changes in natural gas, coal, and oil transportation and delivery networks, establishment of bioenergy crops, decarbonization measures in other industries besides cement and hydrogen production, and efficiency improvements in aviation, rail, and shipping. SOURCE: Committee generated using data from Larson et al. (2020) and Ungar and Nadel (2019).

TABLE 2.2 Comparison of Supply-Side Capital Investment Needed Between 2021 and 2030 in Princeton *Net-Zero America* Study of High Electrification (E+) Net-Zero Pathway and Reference Scenario

Supply Side—Total Capital Investment		Billion USD, 2021–2030	
		Reference	Net-Zero (E+)
Electricity	Wind	110	414
	Solar	62	374
	Natural gas CT and CCGT	101	112
	Natural gas with CCS	0	0
	Li-ion battery storage	3	3
	Biomass with CCS	0	2
Networks	Electricity transmission	203	356
	Electricity distribution	352	369
	EV chargers	1	7
	CO_2 storage	0	11
	CO_2 transportation	0	68
Fuels and industry	H_2—gas reforming	3	3
	H_2—gas reforming with CCS	0	7
	H_2—biomass gasification with CCS	0	0
	Electric boilers	0	12
	Gas boilers	5	5
	Cements with CCS	0	9
	DRI steel	0	0
Total supply side capital expenditure, 2021–2030		840	1,752

NOTE: The Princeton *Net-Zero America* analysis (Larson et al., 2020) quotes both total capital in service for projects that come online from 2021 to 2030 and total capital mobilized, which includes capital being spent in the 2020s for projects that come online post-2030. This table quotes total capital in service. NOTE: CCS = carbon capture and sequestration; DRI = direct reduced iron. SOURCE: Data from Larson et al. (2020).

> **BOX 2.2**
> **MANAGING SYNERGIES, TRADE-OFFS, AND UNINTENDED CONSEQUENCES OF DECARBONIZATION TRANSITIONS**
>
> Carbon, energy, climate, and economic systems are highly integrated with significant positive and negative feedbacks, and complex system effects, both within the United States and worldwide. Relationships and feedbacks among the systems include the type, location, and magnitude of emissions, energy used, material flows, business transactions, and energy services, as well as the resulting climate, health, and economic impacts. These systems will remain highly entwined in a net-zero emissions future, although the magnitude and sometimes direction of the interactions will change. Understanding the current and future relationships and feedbacks among these systems is important for developing effective decarbonization policies and for creating strategies for businesses, organizations, and individuals in response to decarbonization policies.
>
> Policy makers should be attentive to how policies in one sector influence the carbon, energy, climate, social, and economic systems in other sectors. Some actions that reduce emissions and climate damage in one sector are likely to have favorable synergies with decarbonization in other sectors, while others will have negative synergies. Specific areas where interactions may manifest such interactions include land use and the built environment, fuel and material flows, and changes in embodied carbon and life cycle emissions associated with energy end-use technologies.
>
> Policy makers should look to enhance positive synergies while managing negative synergies and unintended consequences. Positive synergies could include decarbonizing both the industrial and transportation sectors with hydrogen, synthetic net-zero carbon fuels, and CCS; facilitating decarbonization of transportation and the built environment through smart growth policies; and developing more efficient energy end-use equipment to provide greater opportunities to manage load and reduce the need for fossil fuels in end-use sectors and electricity generation. Negative synergies can include the need to replace chemical precursors derived from oil refining and the potential impacts on carbon sinks from land conversion to renewable energy production.
>
> Effective management of these synergistic effects would benefit from a whole-economy, comprehensive decarbonization policy process, rather than the current sector-by-sector process governed by congressional committees and agency jurisdiction at the federal level, and similar siloed structures at all levels of government. Both the transition period and the final decarbonized economy will experience new resource needs and constraints, often only emerging after the transition has started. Further research and strategy development in these areas are necessary to minimize negative impacts and create opportunities.

TABLE 2.3 Key Actions Necessary by 2030 for the Five Key Decarbonization Approaches in a Selection of Sectors

	Improve Efficiency and Energy Productivity	Electrify Energy Services in the Buildings, Transportation, and Industrial Sectors	Decarbonize Electricity	Build Critical Infrastructure	Innovate to Complete the Low-Carbon Toolkit
Buildings	• Pursue deep energy efficiency in buildings and appliances. • Work toward achieving carbon-neutral new buildings and 50 percent reduction for existing buildings by 2050.	• After efficiency, update heating and hot water with electric heat pumps to eliminate onsite combustion.	• Integrate flexible demand and onsite energy storage to minimize peak demands. • Integrate photovoltaics on roofs and parking lots where locational value justifies cost.	• Maximize waste heat utilization in district energy systems. • Design waste to energy infrastructures for buildings.	• Reduce carbon in building materials through substitution. • Innovate to sequester carbon in buildings.
Transportation	• Transition to multiple occupancy trips. • Improve ICE fuel efficiency. • Encourage urban planning and infrastructure to facilitate biking and walkability.	• Aggressively pursue zero emission vehicle mandates. • Deploy more zero-carbon transportation fuels.	• Pursue energy efficiency for infrastructure construction and related freight transportation.	• Invest in ubiquitous EV charging infrastructure. • Invest in vehicle connectivity and real-time control infrastructure. • Invest in the H_2 fueling infrastructure, including the design and deployment of fueling stations.	• Reduce costs of renewable fuels. • Repurpose pipeline infrastructure. • Develop improved fuel cells and hydrogen storage.

continued

TABLE 2.3 Continued

Industry	• Triple energy productivity improvement rate. • Expand strategic energy management and smart manufacturing. • Drive system optimization and materials efficiency.	• Develop process heat portfolio, including electricity technology, increasing use 5 percent. • Develop more efficient electrolyzers for H_2, electrochemistry, and direct reduced iron for steel. • Drive early solutions (e.g., hybrid boilers, heat pumps).	• Increase on-site/nearby generation of renewable electricity. • Increase availability of low-carbon electricity at key industrial clusters to catalyze adoption.	• Build capability (substations, HV lines, etc.) to expand use of low-carbon electricity at clusters. • Expand ability to blend H_2 into natural gas networks by upgrading valves, controls, applications. • Build connections between renewable generation and users of H_2.	• Drive RD&D to use green H_2 in processes (e.g., ammonia, MeOH, DRI, high-temperature process heat). • Scale electricity for use in processes (e.g., high-temperature process heat like crackers, or steel making). • Innovative separations technologies that reduce energy use by >50 percent.
Electrical energy storage	• Target widespread adoption of electrified personal and commercial vehicles, buses, trains, and some aircraft.	• Expand adoption of electrified vehicles through lower cost and higher energy density batteries. • Incorporate energy storage with renewable energy generation to enable lower cost industrial operations.	• Where cost effective, install onsite energy storage to accommodate peak usage demands. • Deploy energy storage to facilitate integration of variable renewable energy generation.	• Develop and deploy low-emissions manufacturing and processing of energy storage technologies. • Expand charging infrastructure for electrified vehicles.	• Invest in energy storage RD&D to lower costs, facilitate recycling, adopt low environmental impact materials and processes, increase energy density.

	• Utilize energy storage to enable widespread adoption of renewable energy generation to reduce fossil fuel-burning power plants and peaker plants.	• Where cost effective, deploy energy storage in buildings coupled with use of renewable energy to facilitate electrification.	• Where cost effective, use energy storage to optimize transmission and distribution asset utilization.
Fuels		• Employ electrolysis as a flexible consumer of electricity to produce zero-carbon fuels and feedstocks.	• Expand ability to blend H_2 into natural gas by upgrading valves, controls, and applications. • Build connections between renewable generation and users of H_2.	• Drive RD&D to use green H_2 in fuel formation processes (e.g., ammonia, methanol). • Drive RD&D for air-to-fuels, synfuels, and synthetic aggregate processes with CO_2 as a feedstock.
Electricity generation and transmission		• Increase the share of U.S. electricity from carbon-free sources from ~37 percent of U.S. generation today to roughly 75 percent by 2030.	• Expand long-distance transmission capacity ~120,000 GW-miles by 2030 to connect wind and solar resources to demand centers, a ~60 percent increase.	• Reinforce distribution networks.

continued

89

TABLE 2.3 Continued

			• Enhance the ability of distribution system planning, investment, and operations to use flexible demand and distributed energy resources to improve network asset utilization and efficiently accommodate increased demands from electric vehicles, heat pumps, and other new loads.
		• Deploy ~250–300 GW of new wind and ~280–360 GW of new solar by 2030. • Retire as much as 100 percent of installed coal-fired capacity by 2030 (or retrofit with systems to capture ≥90 percent of CO_2 emissions). • Decrease natural gas-fired generation by ~10–30 percent by 2030 and keep capacity roughly flat nationally	
Carbon capture and sequestration		• Demonstrate and improve use of CCS in methane reforming, biomass gasification, and biofuel production facilities to produce zero-carbon and carbon-negative fuels.	• Increase carbon capture deployment across all sectors for existing technologies. • Plan for pipeline infrastructure required for CO_2 transportation to go online between 2025–2035.	• Drive RD&D for air-to-fuels, synfuels, and synthetic aggregate processes with CO_2 as a feedstock.

- Demonstrate and improve use of bioenergy with CCU to produce CO_2 for Fischer-Tropsch and methanation for production of synthetic drop-in liquid and gaseous fuels.

- Increase deployment of dedicated geologic storage projects in sedimentary basins.

- Improve understanding of CO_2 storage of non-oil and gas reservoirs, including basalts and ultramafic minerals.
- Advance CO_2 mineralization through mine tailings and other industrial alkaline wastes.

NOTE: CCS = carbon capture and sequestration; EV = electric vehicle; ICE = internal combustion engine; RD&D = research, development, and demonstration.

IMPLICATIONS BY SECTOR

The following topical boxes (Boxes 2.3 through 2.9) highlight the committee's evaluation of technologies and approaches required in 2021–2030 to remain on the trajectory for full decarbonization by 2050, organized by sector. Energy demand, supply, carrier, and storage approaches are discussed, including needs for buildings, transportation, industry, energy storage, fuels, electricity generation and transmission, and carbon capture and sequestration. For each topic, the following aspects are highlighted:

- technologies and approaches with the greatest near-term (2021–2030) emissions impact;
- technologies and approaches that have a large potential impact/role in 2031–2050 but need improvement and maturation over the next decade; and
- network infrastructure or other enabling technology or research investment needs that have to be deployed to pave the way to deep decarbonization.

The overall goals for a decarbonization policy plan and the beneficial policies to implement the needed emissions reduction approaches are discussed in Chapters 3 and 4, respectively.

BOX 2.3
BUILDINGS

Building demand reduction presents the largest opportunity to reduce energy demand, as critical to decarbonization as reducing emissions from energy supply. Commercial and residential buildings use 39 percent of total U.S. energy and are responsible for over 35 percent of total U.S. greenhouse gas emissions (EPA, 2020). The built environment can significantly reduce its energy demand, its share of electricity demand, and its embodied carbon. To enable intelligent policy and investment in demand reduction in the building sector, emissions from residential and commercial buildings should be considered together (Figure 2.3.1a) and should consider all associated electricity energy use and emissions (2.3.1b), as well as the embodied carbon in their use of steel, concrete, aluminum, and plastics (2.3.1c). Improvements in the built environment can dramatically reduce energy demand while optimizing asynchronous energy supply (often via thermal storage) and providing measurable gains for productivity, health, and environmental quality.

As evident in the benchmarking data from Seattle displayed in Figure 2.3.2, the worst performing buildings use 2.5–8 times more fossil fuel and electricity than the best performing ones. Demand reductions of 40 percent are easily achievable by 2030, and 80 percent reductions in building energy use intensity (EUI) are achievable by 2050 in the United States, combining new and retrofit construction. Moreover, these massive reductions in demand are some of the most cost-effective investments for decarbonization (McKinsey and Company, 2013).

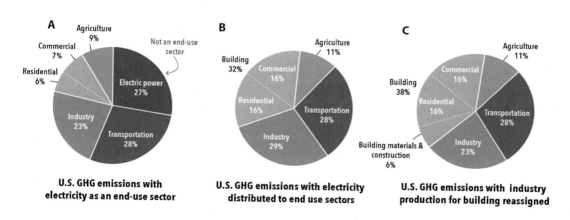

U.S. GHG emissions with electricity as an end-use sector

U.S. GHG emissions with electricity distributed to end use sectors

U.S. GHG emissions with industry production for building reassigned

Fully Assigning GHG Emissions to End Use Sectors for Decarbonization Policy & Action
Created 2020 by Carnegie Mellon Center for Building Performance and Diagnostics, based on Inventory of U.S. Greenhouse Gas Emissions and Sinks 1990-2018, US EPA; Röck at al., 2020

FIGURE 2.3.1a,b,c Collecting all building-related GHG emissions reveals that 38 percent of the environmental challenge is in building construction and operations.
SOURCE: Data from Carnegie Mellon Center for Building Performance and Diagnostics (2020).

continued

BOX 2.3 Continued

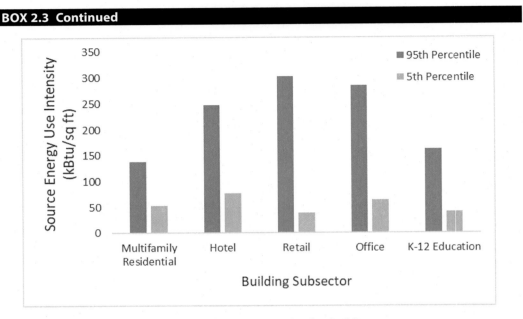

FIGURE 2.3.2 Variation in source energy use intensity (EUI) in five building types.
SOURCE: Committee generated using data from Sullivan (2019).

Six overarching goals and strategies to achieve building demand reduction and decrease carbon emissions from the building sector are described below:

1. **Invest in demand reduction to improve quality of life**, provide U.S. jobs, and reduce inequities. Current U.S. codes, standards, RD&D, and investments in building demand reduction significantly lag behind peer nations. The development of national standards and the removal of market barriers can lead to significant reductions in energy use from key building technologies through their natural replacement cycle. Such standards, which would likely be enforced at a local or state level, are further discussed in Chapter 4.
2. **Make strategic investments in building efficiency and fuel switching** to meet near-term building energy and carbon goals, as outlined in (Ungar and Nadel, 2019):[1]
    ○ *Appliance and equipment efficiency*: 5.6 quads, 210 M $MtCO_2$/yr reductions. Next-generation Energy Star standards and replacements for low-income homeowners offer 70 percent energy savings from a dozen products: residential water heaters, central air conditioners/heat pumps, showerheads, clothes dryers, refrigerators, faucets, and furnaces, as well as commercial/industrial fans, electric motors, transformers, air compressors, and packaged unitary air conditioners and heat pumps.

BOX 2.3 Continued

- Net-zero emissions in new homes and commercial buildings: 5.7 quads, 265 M MtCO$_2$/yr reductions. Standards and low-income homeowner incentives offer 70 percent energy savings relative to reference-case efficiency levels, with the remaining 30 percent coming from on-site or off-site carbon-free energy systems.
- Smart homes and commercial buildings—new and existing: 3.2 quads, 125 M MtCO$_2$/yr reductions. Weatherization Assistance Program (WAP) training and employment for smart controls, access to real-time information, and smart algorithms will optimize energy savings for automation systems in both residential and commercial buildings (Elliott et al., 2012).
- District and combined heat, cooling, and power—new and existing: 4 quads, 150 MtCO$_2$/yr reductions. Co- or poly-generation of power, heating, hot water, and cooling with district energy systems can reduce emissions by 150 million metric tons of CO$_2$ each year (M MtCO$_2$/yr) by installing new combined heat and power (CHP) plants with a total capacity of 40 GW by 2020 (Park et al., 2019). As long as there is sufficient waste heat from industry and power generation (including increases in waste-to-power), district energy systems offer substantial efficiencies in mixed-use communities in heating dominated climates and offer resiliency for hospitals, schools, and community spaces.
- Existing home and commercial building envelope retrofits: 3.8 quads, 125 M MtCO$_2$/yr reductions. WAP training and employment for retrofits that improve air tightness, envelope insulation, and window quality to meet ENERGY STAR can reduce energy use by 20–30 percent and improve comfort and health (Belzer et al., 2007; Liaukus, 2014). All commercial buildings undergoing major retrofits should achieve 50 percent reductions in demand (Shonder, 2014).
- Electrification of space heating and water heating in existing homes and commercial buildings: 0.9 quads (after measures above), 76 M MtCO$_2$/yr reductions. Industry standards and incentives can accelerate the deployment of high-efficiency heat pumps that use electricity from low- or no-carbon generation, including on-site photovoltaics that can offer a level of resiliency.

3. **Reduce embodied carbon emissions.** As buildings become more efficient, the embodied carbon in building materials becomes as critical as operational carbon. The embodied carbon emissions from all new buildings, infrastructures, and associated materials should be reduced by 50 percent by 2030 and eliminated by 2050.
4. **Electrify the built environment and integrate it with the grid.** Buildings have a role in electricity generation, storage, and carbon sequestration as well. Buildings and communities play a significant role in decarbonizing energy supply through the following:
   - Electrification of the built environment with the lowest conditioning, process, plug and parasitic loads through conservation, passive conditioning, and energy cascades;
   - Peak load shaving and demand flexibility;
   - District and building CHP for 150 M MtCO$_2$/yr;

continued

BOX 2.3 Continued

- ○ Site- and building-integrated photovoltaics and solar thermal, where cost-effective;
- ○ Thermal energy storage (water, ice, phase change materials);
- ○ Geothermal, aqua-thermal, and ground-coupled HVAC; and
- ○ Site-generated electricity and off-peak electricity storage.

5. **Enhance the carbon sequestration ability of buildings and infrastructures through a series of innovations:**
   - ○ Increasing the use of wood construction from sustainably harvested forests (SFC) to reduce or replace steel, aluminum, and concrete;
   - ○ Encapsulating CO_2 into aggregate and/or the sand that makes up 85 percent of concrete to sequester up to 1,200 pounds of CO_2 per cubic yard of concrete and allow buildings to be carbon negative; and
   - ○ Restoring indigenous landscapes through green roofs and the reforestation of urban, suburban, and rural communities.

6. **Adopt the New Buildings Institute's five foundations of Zero-Carbon Building Policy:** energy efficiency, renewable energy, grid integration and storage, building electrification, and embodied carbon. A net-zero carbon or a net-negative carbon built environment is key for the decarbonization of the United States. Energy Use Intensities should be driven by code to achieve passive house standards of less than 25 to 50 kBTU/sqft per year depending on building type. This should be followed by integrating site and community renewable energy sources with effective grid integration and energy storage, wherever cost effective. These actions should fully anticipate the elimination of fossil fuels and combustion in buildings, with building electrification as a linchpin solution for decarbonization of the United States. Last, the built environment offers a path to carbon sequestration, with sustainably managed forests and the use of carbon sequestering materials. The optimum mix of investments in design for deep efficiencies, electrification with site and community generation, and reduced carbon in building material production or even carbon sequestering material can ensure that the building sector achieves net positive in carbon sequestration (Webster et al., 2020).

---

[1] The CO_2 savings reported for each efficiency and fuel switching investment reflect current grid and fuel emissions levels.

BOX 2.4
TRANSPORTATION

In 2018, transportation carbon emissions were predominantly from roadway vehicles, including light-duty vehicles (59 percent) and medium- and heavy-duty trucks (24 percent) (EPA, 2020). Aircraft contributed 10 percent, pipelines emitted 3 percent, while ships and rail each contributed 2 percent. Because vehicles last a decade or more (see Figure 2.4), the next decade should prepare the United States for a major change in vehicle fleet emissions. Priorities and opportunities in the transportation sector are outlined in this box.

1. **Improve efficiency and energy productivity.** Numerous policies and actions can improve transportation efficiency and energy productivity. Avoiding travel can reduce energy use by 10–30 percent (Lah, 2017; Ungar and Nadel, 2019) with teleworking, encouraging compact, mixed-use cities, local production of food and goods, and others. Shifting travel to more energy efficient modes can also reduce energy use by 10–30 percent (Lah, 2017; Ungar and Nadel, 2019). Examples include vehicular ride sharing (including mass transit), using rail rather than trucking, or adopting biking and walking. Improving the performance of vehicles and transportation networks can also save energy. Fuel efficiency of new light-duty vehicles has doubled over the past 30 years. Data analytics and improved communications permit better management of roadway networks. Safer automated vehicles can reduce crashes and congestion associated with crashes.

2. **Electrify Vehicles.** Both battery electric and fuel cell vehicles are now offered for commercial sale in the United States. In 2018, 240,000 battery electric vehicles were sold in the United States, representing 1.4 percent of all light-duty vehicle sales (Davis and Boundy, 2020). Battery electric buses and fuel cell vehicles are also available, but sales are much smaller than for light-duty battery electric vehicles. Fuel cell vehicles still have a sizable cost premium and limited hydrogen filling stations. With the continued improvement in battery performance, the extra cost of battery electric vehicles is expected to be small by 2025 (Lutsey and Nicholas, 2019), while the operating costs will likely be lower than conventional vehicles. At the same time, the range of battery electric vehicles has been increasing, with some commercial vehicles offering a 250-mile range or greater. Further, light-duty trucks and buses should be electrified, particularly in urban areas. Over the next decade, the United States needs to ensure that electric vehicles become the predominant share of new purchases.

   Infrastructure investment is required to enable this switch to electrification. Vehicle charging or hydrogen fueling stations must become widely available. Railroad catenary infrastructure can extend electric locomotive use. Ships and aircraft should switch to grid connections while in port or taxiing.

continued

BOX 2.4 Continued

Widespread vehicle electrification should also provide supply chain manufacturing and service opportunities. For example, the United States could become a leader in battery manufacturing and innovation. Vehicular maintenance and training would need to change.

3. **Other Actions.** Some aircraft services, such as package delivery via drones, could be electrified. Similarly, ships could use fuel cells or small nuclear reactors as power sources. Nevertheless, liquid fuels may be the most cost-effective fuel for long-haul transportation services. Research and development is needed to reduce the cost of producing low-carbon synthetic liquid transportation fuels (discussed further in Box 2.5). Pipelines could be repurposed for carbon dioxide transport and powered by electricity rather than by fossil fuels.

BOX 2.5
INDUSTRY

The U.S. industrial sector is crucial for GHG reductions, accounting for 32 percent of the nation's energy use including feedstocks (EIA, 2018), 22 percent of GHG emissions, and around a billion metric tons of CO_2 emissions/year (EPA, 2020). The sector is an important part of the U.S. economy, accounting for 11 percent of the gross domestic product (GDP) and 13 million direct jobs (NAM, 2018). Thus, it is vital to pursue the transformation while safeguarding competitiveness.

Energy inputs for manufacturing are 83 percent from hydrocarbons and 17 percent from electricity (EIA, 2014). The industrial sector is diverse, complex, and intertwined with multilevel value chains. Refining, chemicals, iron and steel, food products, and cement account for the largest portion of the energy use and CO_2 emissions. Feedstocks are an important source of embedded energy in chemical manufacture, where they account for up to 60 percent of the combined energy. Process heat uses 61 percent of the on-site energy accounting for 32 percent of GHG emissions and 7 percent of GHG emissions across sectors, with 90 percent from fossil fuels (DOE, 2015c; EIA, 2020e).

Given the variation of energy sources, multiple uses, diverse product mix, reliance on carbon for products, and variation in the regional-grid GHG emission intensities, it will be critical to proactively pursue multiple decarbonization pillars in parallel. Low-carbon technologies, approaches, and infrastructure needing RD&D investment are shown in Figure 2.5.1.

Cross-cutting opportunities across sectors include process heat, switching to low-carbon energy sources, separations, electrolyzer efficiency, motor efficiency, and recycling. Sector specific opportunities abound, including transformative process technologies, renewable H_2 use in processes, and thermal transfer.

BOX 2.5 Continued

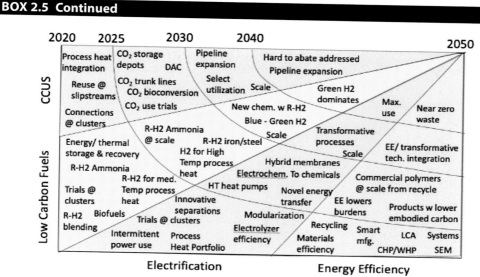

FIGURE 2.5.1 RD&D investment needs for low-carbon technologies, 2020–2050.

During the next 10 years, the key strategies to rapidly and persistently pursue are as follows:

1. **Energy efficiency.** Accelerate low-capital solutions (e.g., energy, materials, system efficiency; separations, intermittent fuel switching), greatly expand strategic energy management and smart manufacturing.
2. **Electrification.** Develop and deploy a process heat solutions portfolio featuring electric technologies and more effective electrolyzers for H_2. Drive RD&D on electrified processes. Build infrastructure to deliver low-carbon electricity to industrial facilities reliably.
3. **Hydrogen/low-carbon energy and feedstocks.** Rapidly trial and scale up the use of zero-carbon H_2 via blending, in transformative processes at clusters, and for high-temperature process heat. Advance RD&D in these areas.
4. **CCUS.** Integrate for lower costs, expand infrastructure starting at clusters, expand CO_2 utilization.
5. **Anticipate/minimize trade-offs.** For example, benzene, toluene, xylene (BTX) dependence on refineries, H_2 moisture, missing by-product.

BOX 2.6
ELECTRICAL ENERGY STORAGE

Reports demonstrate that the greatest reductions in GHG emissions in the near term (2021–2030) are achievable through (1) electrification of end uses and (2) decarbonization of electricity generation (Williams et al., 2014; Haley et al., 2019; Larson et al., 2020). Electrical energy storage will play an important role for both objectives with major impact in transportation and electricity supply.

Transportation accounts for roughly 30 percent of all greenhouse gas emission in the United States, with much of that attributed to personal vehicles. Currently, a primary barrier to the widespread adoption of EVs is cost. Battery materials account for 60 percent of the battery cost and therefore are of keen importance for research and development (Leuenberger and Frischknecht, 2010). Although lithium composes only 2 percent of the total battery cost, its supply chain availability is at risk (Majeau-Bettez et al., 2011; BNEF, 2019; Harper et al., 2019). Modifications to the cathode manufacturing process and introduction of recycling/regeneration approaches may aid in the continual cost reduction (Poyraz et al., 2016; Harper et al., 2019; Wood et al., 2020). A secondary issue limiting adoption is "range anxiety," which could be addressed with a dual strategy of higher energy density batteries as well as fast charge technology (Li et al., 2001; Tallman et al., 2019; Zheng et al., 2019; IEA, 2020b; Woo and Magee, 2020). An emerging direction is adoption of conversion or alloying electroactive materials that enable multiple electron transfers per active center, dramatically increasing the battery energy density. Resolution of the long-term stability of these systems would enable their adoption. Lithium-ion battery researchers are demonstrating that at the cell level, charge in ~10 minutes may be viable and should be pursued (Ahmadi, 2019; Tallman et al., 2019; Chen et al., 2020). Fast charging stations should be accompanied by renewable energy generation where large loads can be offset through local energy storage and the likelihood of >1 MW being drawn from the grid at once is mitigated (Bhatti et al., 2016). Electrification of light commercial vehicles must also expand. In order to enable the needed >400-mile range, a specific energy of >200 Wh/kg at the pack level must be achieved and could be realized through next generation lithium metal/sulfur and lithium/air batteries.

As renewable energy becomes an increasing part of the U.S. power grid, the need for flexible energy storage increases. Currently, introducing renewables into the U.S. grid infrastructure can increase grid variability owing to inherent intermittency and will increase the discrepancy between electricity supply and demand. Implementation of energy storage technologies can mitigate these issues using electrochemical (batteries, redox flow batteries, supercapacitors, and fuel cells) or non-electrochemical (pumped hydro, compressed air, thermal, flywheel, and superconducting magnetic) approaches. The selection of the appropriate storage is dictated by factors including location, power demand, discharge time, and cost. Notably, the locational limitations can be significant, particularly in urban areas; thus, the discussion here is focused on energy storage with locational flexibility. Currently, >80 percent of large-scale battery storage capacity is from Li-ion batteries (EIA, 2018). Despite their availability and widespread deployment, there are safety and cost concerns associated with introducing Li-ion batteries at the grid level (DOE, 2014; Balaraman, 2020). Expanded research and development with subsequent deployment of batteries or redox flow batteries with aqueous electrolytes can provide safer, lower cost, lower environmental impact, and more scalable alternatives to the nonaqueous electrolytes currently

BOX 2.6 Continued

used in Li-ion technologies (DOE, 2018; Wang et al., 2019; Kim et al., 2020). Further, assessing battery aging and failure mechanisms for candidate battery types is an ongoing and important area of inquiry related to prediction of deployment lifetime and the associated costs of installation and replacement (Palacín et al., 2016). Added efficiency may be possible through coupling of electrochemical energy storage with other methods of storage such as thermal, compressed air, or flywheel still providing installation flexibility. Energy storage technologies capable of achieving very low cost per kWh of storage capacity (on the order of $1–$5/kWh; Sepulveda et al., 2021) may ultimately serve as long-duration electricity storage technologies capable of addressing intermittency over weeks-long time periods. Advanced electrochemical, chemical, and thermal storage technologies would be needed to serve in this role and their design and capabilities are very different from shorter-duration grid-scale storage applications such as Li-ion batteries or conventional flow batteries.

Key strategies to address the above challenges include

1. **Reduce cost of batteries for transportation,** including consideration of factors related to materials selection, supply chain, regeneration, and recycling.
2. **Increase energy density and develop fast charge capability** of batteries to enable expanded adoption of electrified passenger and commercial vehicles as well as some aircraft.
3. **Develop low-cost, environmentally benign, safe, long-life electrochemical energy storage** for use with renewable energy generation to provide flexibility of site selection, including a range of designs and capabilities suitable for cycling over intraday, interday, and weekly time periods or longer.
4. **Couple electrochemical energy storage with thermal and mechanical methods,** when possible, to gain efficiency and extend total storage time.

> **BOX 2.7**
> **FUELS**
>
> Fuel-based energy carriers are deeply embedded in society, have a major infrastructure base, and have very large power densities. As such, even while electrification is important, it is unlikely to be the exclusive approach to enable decarbonization, particularly over the 2050 time frame. In addition, some sectors are harder than others to electrify, and net-zero carbon chemical energy carriers will likely remain the lowest cost option for certain sectors such as aviation and shipping. From a broader perspective, developed economies already have massive built out fuels-handling, logistical, and midstream infrastructures that can be leveraged immediately. Legacy equipment can be readily decarbonized if the fuel is decarbonized.
>
> A variety of chemical energy carriers can be produced with net-zero CO_2 emissions; estimated current costs are summarized in Table 2.7.1. A convenient way to organize these options
>
> **TABLE 2.7.1** Comparison of Costs on an Energy Basis for Various Energy Carriers
>
> | Energy Carrier | ($/Gigajoule)[a] |
> |---|---|
> | Conventional natural gas[b] | 3 |
> | Conventional industrial H_2 from natural gas[c] | 7 |
> | Conventional gasoline[d] | 15 |
> | Renewable hydrogen from electrolysis[e] | 35 |
> | Renewable CO_2 gasoline[f] | 55 |
> | Renewable ethanol fuel[g] | 16 |
> | Ammonia from methane[h] | 22 |
> | Renewable ammonia[i] | 30 |
>
> [a] Note that these numbers come from different sources with different assumptions and so provide general guidance on pricing, but all can move up or down based on assumptions (e.g., electricity prices). For example, this is likely the reason that the hydrogen production cost is slightly higher than the ammonia.
> [b] Average Henry Hub Price 2018 from EIA, 2019b.
> [c] $1/kg price 2018 estimate reported in Bonner, 2013.
> [d] Spot price for RBOB gasoline in 2019 from Investing.com, 2019.
> [e] $5/kg based on estimate reported in IRENA, 2018.
> [f] $200/MWh fuel based on Brynolf et al., 2018.
> [g] Price from EIA, 2019c and at 89 MJ/gallon.
> [h] Price of $500/ton as reported in Schnitkey, 2018.
> [i] Price-based factor from Schnitkey, 2018 and EPRI, 2019.
> SOURCE: Courtesy of Prof. Matthew Realff, Georgia Institute of Technology.

BOX 2.7 Continued

is (1) if they emit carbon when oxidized and (2) if they can "drop-in," without requiring changes to the existing distribution infrastructure and users. Non-drop-in options, like hydrogen, ammonia, or ethanol, can be inserted into current systems as mixes with drop-in fuels. For example, the existing fleet of natural gas-fired power plants can operate with hydrogen levels of up to about 5 percent, and automobiles with ethanol levels of 15 percent. However, such energy carriers cannot be used in significant concentrations without modifying users/carrier infrastructure.

Research and demonstrations should continue for both drop-in and non-drop-in options. The following briefly highlights priority research needs for high-potential fuels in each category.

1. **Reduce the cost of low-carbon hydrogen production methods and technologies that utilize hydrogen** (e.g., fuel cells). Hydrogen (H_2) is a non-drop-in fuel and the lowest cost synthetic fuel per energy content. It can be produced by the electrolysis of water using renewable energy sources, via steam methane reforming and autothermal reforming of natural gas (including with CCS, rendering the process zero- or near-zero carbon), and from biomass gasification (including with CCS for a net negative emissions process).

2. **Minimize the costs and/or maximize utilization of the existing fuel infrastructure,** such as increasing the hydrogen level that gas turbines can accommodate. Maximum insertion levels of H_2 into existing transportation systems and consuming devices are set by user requirements (e.g., premixed versus nonpremixed gas turbines or heaters) and pipeline embrittlement concerns. H_2 injection into the existing pipeline at low, but progressively increasing, levels is an example of a needed demonstration project.

3. **Fund R&D on enabling technologies for low-carbon synthetic fuel production.** A variety of synthetic fuels, including ethanol, methane, and gasoline or aviation gas substitutes, emit carbon when combusted. If they are "drop-in" substitutes, such energy carriers would use the existing hydrocarbon infrastructure. To make them net-zero carbon, these drop-in fuels must be produced from captured carbon dioxide and zero-carbon hydrogen and can be synthesized via existing chemical processes, such as water gas shift and commercially available Fischer-Tropsch chemistry. The synthesis of low-carbon liquid fuels is very energy intensive and will require significant amounts of clean energy, likely electricity. Another pathway is through photosynthetic conversion of CO_2 and water into biomass and then its subsequent treatment to produce a drop-in fuel. Low-carbon synthetic fuels are currently more expensive to produce than hydrogen—as such, it is important for research and demonstration projects to prioritize driving down costs.

BOX 2.8
ELECTRICITY GENERATION AND TRANSMISSION

The electricity sector is a linchpin in any successful transition to a net-zero emissions U.S. economy by 2050 or sooner (Jenkins et al., 2018; Haley et al., 2019; Larson et al., 2020). Pathways to cost-effectively reach net-zero greenhouse emissions entail twin challenges for the electricity sector:

- As the source of more than a quarter of U.S. greenhouse gas emissions and with multiple scalable, affordable alternatives to fossil fueled power plants available today, the electricity sector must (and can) cut emissions faster and deeper than any other sector (Phadke et al., 2020; Haley et al., 2019; Vibrant Clean Energy, 2019).
- *Electricity generation must substantially expand*—approximately 10–20 percent by 2030 and 120–170 percent by 2050—to fuel a greater share of energy use in transportation, building space heating, and low- and medium-temperature industrial process heat as well as produce hydrogen from electrolysis and even power direct air capture (Larson et al., 2020).

Reducing power-sector emissions rapidly toward zero while expanding electricity production involves the following key strategies:

1. **Expansion of carbon-free electricity generation.** The share of U.S. electricity from carbon-free sources roughly doubles from about 37 percent of U.S. generation today to roughly 75 percent by 2030 (Larson et al., 2020; Phadke et al., 2020) and ~100 percent by 2050 or sooner. As electricity demand grows during this period, this entails bringing online roughly 2 billion MWh of new carbon-free generation by 2030, enough to supply about half of current U.S. electricity production (EIA, 2020a).
2. **Wind and solar power.** Wind and solar power capacity expands rapidly, with ~250–300 GW of new wind (2–3 times existing capacity) and ~280–360 GW of new solar (~4 times existing capacity) deployed by 2030 (DOE-EERE, 2020a; SEIA, 2020; Larson et al., 2020). By this date, wind and solar supply about 45–55 percent of electricity nationwide (up from 10 percent today).
3. **Coal power.** As much as 100 percent of installed coal-fired capacity retires by 2030 (or is retrofit with systems to capture 90 percent or more of CO_2 emissions). Phasing out (or capturing CO_2 from) coal-fired power reduces U.S. CO_2 emissions by ~1 Gt/year, one-sixth of total net U.S. GHG emissions, while avoiding approximately 40,000 deaths and $400 billion in air pollution damages during 2021–2030 (EIA, 2020a; EPA, 2020; Larson et al., 2020).
4. **Nuclear power.** Existing U.S. nuclear plants provide nearly 100 GW of firm low-carbon generation capacity and supply almost one-fifth of U.S. electricity today (EIA, 2020a). Preserving existing reactors, wherever they are safe to continue operating, provides a cost-effective foundation of low-carbon electricity to build toward decarbonization goals.

BOX 2.8 Continued

5. **Natural gas power.** Natural gas-fired generation declines modestly by ~10–30 percent by 2030 and installed capacity is roughly flat nationally. Remaining gas plants play a key role providing "firm" capacity that is available on demand for as long as needed without dependence on weather (Sepulveda et al., 2018). Existing natural gas capacity may be maintained through 2050 to provide firm capacity, if operated much less frequently and if hydrogen is blended with (or replaces) natural gas to fuel these plants when they do operate to reduce (or eliminate) carbon emissions intensity.
6. **Energy storage.** ~10–60 GW / 40–400 GWh of battery energy storage capacity is likely needed through 2030 (Larson et al., 2020; Phadke et al., 2020). Storage will likely play a larger role from 2030–2050; during this period, production of hydrogen, ammonia, or synthetic methane from carbon-free electricity can offer a form of longer-term chemical energy storage, and these energy carriers can be used as fuels (or feedstocks) in transportation, heating, and industry or for firm power generation.
7. **Transmission networks.** Long-distance transmission capacity expands ~120,000 GW-miles by 2030 to connect wind and solar resources to demand centers, a ~60 percent increase (Larson et al., 2020; Hand et al., 2012). By 2050, long-distance transmission capacity may need to more than triple as electricity demand grows and renewable resources play a central role in the U.S. grid (Larson et al., 2020).
8. **Distribution networks.** Distribution networks are reinforced and distribution system planning, investment, and operations make better use of flexible demand and distributed energy resources to improve network asset utilization and efficiently accommodate up to 10 percent increase in aggregate peak demands from electric vehicles, heat pumps, and other new loads; prepare for up to 40 percent increase in peak demands by 2040 and roughly 60 percent increase by 2050.

BOX 2.9
CARBON CAPTURE AND SEQUESTRATION

Capturing CO_2 from point sources avoids emissions into the atmosphere and reduces the carbon footprints of electricity and industrial products such as cement, steel, bioethanol, and so on.

CCS is more likely to be used to decarbonize gas power plants, rather than coal power plants. Over 50 percent of today's coal fleet is over the typical retirement age of 40 years (EIA, 2019a). On the other hand, the fleet of natural gas power plants is relatively young; the majority reach retirement between 2030–2050 (Figure 2.9.1). Natural gas units are therefore more suitable for CCS, which could avoid up to ~700 $MtCO_2$/yr at a cost today of ~ $60/$tCO_2$ avoided (Psarras et al., 2020). The bar graph inset (see Figure 2.9.1) indicates retirement period of the coal and natural gas power fleets along with the number of units (above bars) and cumulative capacities assuming an average age of 40 years. Biomass power plants with CCS could also be deployed in the future to produce electricity with net negative CO_2 emissions.

In the industrial sector, ~330 $MtCO_2$/yr of current process emissions are suitable for carbon capture retrofit (Pilorgé et al., 2020). A number of industries produce CO_2 as a chemical by-product of their industrial process in relatively high CO_2 concentrations, making the separation of CO_2 easier and less costly than more dilute streams of combustion exhaust. The relative scales and exhaust stream concentrations are shown in Table 2.9.1, along with their nth-of-a-kind cost estimates (NASEM, 2019; McQueen et al., 2020; Pilorgé et al., 2020; Psarras et al., 2020). By 2050, production of hydrogen, drop-in fuels and feedstocks from biomass waste with CCS could present a significant

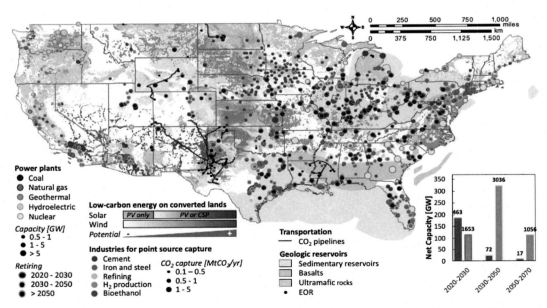

FIGURE 2.9.1 Existing CO_2 sources from power and industrial sectors overlaid with renewable energy potential, in addition to CO_2 storage opportunities, and existing and potential low-carbon power plants that would be coupled to CO_2 storage.

BOX 2.9 Continued

opportunity as well, delivering roughly 300–1,000 MtCO$_2$/yr of negative emissions by 2050 (Larson et al., 2020) that can offset direct emissions elsewhere in hard-to-decarbonize sectors.

Avoiding CO$_2$ emissions through capture at the point source requires less energy than capture from dilute air. However, direct air capture is receiving increased attention as an approach to offset difficult-to-eliminate emissions from the transportation, industrial heating, and agriculture sectors, and for other options in carbon removal (Wilcox, 2020a,b).

TABLE 2.9.1 Scale, Energy, Cost, and Example Carbon Capture Projects Globally

Capture Application	2020-Scale (MtCO$_2$/yr)[a]	Percent CO$_2$[b]	Min Work (kJ/mol)[c]	Nth-kind Cost[d] ($/tCO$_2$)	Example Projects (Start Date; Scale)
Natural Gas	700	3–5	~9–10	~55–60	Elk Hills, Fluor (2020; Mt/yr)
Industry (process emissions only)					
Cement	67	25–30	~4	~30	Lafarge Holcim, Total, Svante (2019; kt/yr)
Refining	40	15–20	~6	~40	Norway, Statoil Mongstad (2012; 100s kt/yr)
Bioethanol	37	99+	~0	~<20	Decator, ADM (2017; Mt/yr)
Hydrogen	26	45–70	~2–3	~25–30	Port Arthur, Air Products (2013; Mt/yr)
Iron and Steel	19	20–25	~5	~35–40	Abu Dhabi CCS Project, UAE (2016; Mt/yr)
Air Capture[e]					
Solvents	~1	0.04	21	~150–600	Carbon Engineering (2023; 1 Mt/yr)
Solid Sorbents	<1	0.04	21	~150–600	Climeworks—14 plants globally (kt/yr)

[a] Pilorgé et al., 2020; Psarras et al., 2020.
[b] Bains et al., 2017; Pilorgé et al., 2020; Psarras et al., 2020.
[c] Wilcox, 2012.
[d] Based on nth-of-a-kind technology-agnostic modeling from Pilorgé et al., 2020.
[e] Cost range is broad and depends on technology and energy resource choices, in addition to scale of deployment.

continued

BOX 2.9 Continued

Technologies exist for all applications of carbon capture, but in many cases have not yet achieved the scale, maturity, or cost required for the decarbonization needed (e.g., $GtCO_2/yr$). For instance, solid sorbent-based technologies such as those developed by Svante (Global Cement, 2019, 2020) are modular approaches that may be optimized and scaled across the industrial sector where emissions ($ktCO_2/yr$) and CO_2 concentrations vary widely. For the power sector, owing to sizable emissions per facility (i.e., $MtCO_2/yr$), separation processes with significant economies of scale, such as solvents (e.g., Fluor) (Reddy and Freeman, 2018), are a good match for retrofitting natural gas plants retiring beyond midcentury.

Key strategies for CCS in the next decade are as follows:

1. **Increase deployment of all carbon capture technologies** shown in Table 2.9.1 over the next decade so that decarbonization can reach the 100 $MtCO_2/yr$ scale and DAC deployment occurs on the $MtCO_2/yr$ scale. This will require a roughly 10× increase in carbon capture activities.
2. **Develop permanent storage of CO_2 on the Gt-scale.** This amount of storage will be required regardless of the source of CO_2, and Figure 2.9.1 shows a number of such opportunities, from geologic storage in saline aquifers to depleted oil and gas reservoirs, to the more nascent opportunities such as alkaline-rich formations like basalts and ultramafic rocks (NASEM, 2019; Kelemen et al., 2019).

CONCLUSION

The many feasible pathways for deep decarbonization by 2050 all have similar requirements in 2021–2030, which are summarized in this report's five key actions for the 2020s:

1. Invest in energy efficiency and productivity across end-use sectors.
2. Electrify energy services, particularly transportation and heating.
3. Produce carbon-free electricity, doubling the share of clean electricity generation to roughly 75 percent.
4. Plan, permit, and build critical infrastructure and repurpose existing energy infrastructure.
5. Expand the innovation toolkit, investing in RD&D and creating initial markets for nascent net-zero technologies via incentives and standards.

In order to remain on a path to decarbonization by 2050, decision makers in transportation, buildings, industry, electricity generation/transmission/distribution, fuels, and other sectors must align their actions with the five key actions. Some actions for 2021–2030 are already under way with existing technology and need to be maintained or accelerated, such as the steady installation of new zero-carbon electricity generation and retirement of coal-fired generation. Some actions could be implemented immediately with existing technology but need to begin, such as replacement of building heating with electric equipment, widespread deployment of electric vehicles, and blending of hydrogen with natural gas in industrial infrastructure and equipment. Other actions, such as carbon capture, require improvement and maturation of existing technologies or new technology or approaches to be developed and tested at scale, and research, development, and deployment in these areas must be accelerated in the decade after the release of this report to provide options in 2030–2050.

The investment requirements in 2021–2030 to implement the five key actions would require spending no more of the nation's GDP on energy services than the United States has over the past decade and would require a total incremental expenditure of about $300 billion above the business-as-usual case through 2030 (an ~3 percent increase relative to business-as-usual). The transition to a decarbonized system would have significant benefits in the United States, on the order of $200 billion to $300 billion annually of avoided climate damages, in addition to preventing hundreds of thousands of premature deaths and saving trillions of dollars of health costs from fossil fuel pollution.

Although the approaches to decarbonization are well known, and the technologies to get started are ready for implementation in 2021–2030, new policies and systems are required to reduce cost and encourage adoption of needed technologies and approaches at a sufficient pace and scale. The technological goals for a net-zero energy system by 2050 detailed in this chapter are complementary to the socioeconomic goals discussed in Chapter 3 and integral to the policy options presented in Chapter 4.

REFERENCES

Ahmadi, P. 2019. Environmental impacts and behavioral drivers of deep decarbonization for transportation through electric vehicles. *Journal of Cleaner Production* 225: 1209–1219.

Architecture 2030. n.d. "The 2030 Challenge." https://architecture2030.org/2030_challenges/2030-challenge/.

Bains, P., P. Psarras, and J. Wilcox. 2017. CO_2 capture from the industry sector. *Progress in Energy and Combustion Science* 63: 146–172.

Balaraman, K. 2020. "Why Is the Utility Industry Less Bullish on Grid-Scale Storage?" Utility Dive. https://www.utilitydive.com/news/safety-volatile-market-less-bullish-storage/572013/.

Bataille, C., H. Waisman, M. Colombier, L. Segafredo, and J. Williams. 2016. The Deep Decarbonization Pathways Project (DDPP): Insights and emerging issues. *Climate Policy* 16(Suppl 1) S1–S6.

Belzer, D., G. Mosey, P. Plympton, and L. Dagher. 2007. *Home Performance with ENERGY STAR: Utility Bill Analysis on Homes Participating in Austin Energy's Program*. National Renewable Energy Laboratory. https://www.osti.gov/biblio/910503.

Bhatti, A.R., Z. Salam, M.J. Bin Abdul Aziz, K.P. Yee, and R.H. Ashique. 2016. Electric vehicles charging using photovoltaic: Status and technological review. *Renewable and Sustainable Energy Reviews* 54: 34c47.

BNEF (Bloomberg New Energy Finance). 2019. "New Energy Outlook 2019." *BloombergNEF* (blog). https://about.bnef.com/new-energy-outlook/.

Bonner, B. 2013. "Current Hydrogen Cost." Presented at the DOE Hydrogen and Fuel Cell Technical Advisory Committee, October 30. https://www.hydrogen.energy.gov/pdfs/htac_oct13_10_bonner.pdf.

Brown, D.R. 2015. Hydrogen production and consumption in the U.S.; The last 25 years. *Cryogas International* 53(9):40–41.

Brynolf, S., M. Taljegard, M. Grahn, and J. Hansson. 2018. Electrofuels for the transport sector: A review of production costs. *Renewable and Sustainable Energy Reviews* 81: 1887–1905.

Chen, K., M.J. Namkoong, V. Goel, C. Yang, S. Kazemiabnavi, S.M. Mortuza, E. Kazyak, et al. 2020. Efficient fast-charging of lithium-ion batteries enabled by laser-patterned three-dimensional graphite anode architectures. *Journal of Power Sources* 471.

Cox, S., S. Tegen, I. Baring-Gould, F.A. Oteri, S. Esterly, T. Forsyth, and R. Baranowski. 2015. *Policies to Support Wind Power Deployment: Key Considerations and Good Practices*. Golden, CO: National Renewable Energy Laboratory.

Davis, S., and R.G. Boundy. 2020. *Transportation Energy Data Book: Edition 38*. Oak Ridge, TN: Oak Ridge National Laboratory.

Davis, S.J., N.S. Lewis, M. Shaner, S. Aggarwal, D. Arent, I.L. Azevedo, S.M. Benson, et al. 2018. Net-zero emissions energy systems. *Science* 360(1419).

de Pee, A., D. Pinner, O. Roelofsen, K. Somers, E. Speelman, and M. Witteveen. 2018. *Decarbonization of Industrial Sectors: The next Frontier*. Washington, DC: McKinsey and Company.

DOE (Department of Energy). 2014. *Energy Storage Safety Strategic Plan*. Washington DC: Office of Electricity Delivery and Energy Reliability.

DOE. 2015a. "Revolution . . . Now: The Future Arrives for Five Clean Energy Technologies—2015 Update." https://www.energy.gov/sites/prod/files/2015/11/f27/Revolution-Now-11132015.pdf.

DOE. 2015b. *Guide to Determining Climate Regions by County*. Building American Best Practices Volume 7.3. Richland, WA: Pacific Northwest National Laboratory.

DOE. 2015c. *2015 Improving Process Heating System Performance: A Sourcebook for Industry, Third Edition*. Washington DC: Office of Energy Efficiency and Renewable Energy.

DOE. 2018. *Spotlight: Solving Challenges in Energy Storage*. Washington, DC: Office of Technology Transitions.

DOE-EERE (Department of Energy Office of Energy Efficiency and Renewable Energy). 2020a. "WINDExchange: U.S. Installed and Potential Wind Power Capacity and Generation." https://windexchange.energy.gov/maps-data/321.

DOE-EERE. 2020b. "Advancing the Growth of the U.S. Wind Industry: Federal Incentives, Funding, and Partnership Opportunities." Washington DC. https://www.energy.gov/sites/prod/files/2020/02/f71/weto-funding-factsheet-2020.pdf.

EFI (Energy Futures Initiative). 2019. *The Green Real Deal: A Framework for Achieving a Deeply Decarbonized Economy.* Washington DC.

EIA (Energy Information Administration). 2014. "Table 3.2 Energy Consumption as a Fuel by Mfg. Industry and Region (Trillion BTU)." 2014 Manufacturing Energy Consumption Survey (MECS) Survey Data. https://www.eia.gov/consumption/manufacturing/data/2014/pdf/table3_2.pdf.

EIA. 2018. "U.S. Battery Storage Market Trends." May 21. https://www.eia.gov/analysis/studies/electricity/batterystorage/archive/2018/.

EIA. 2019a. *Annual Energy Outlook 2019 with Projections to 2050.* January 29. Washington, DC. https://www.eia.gov/outlooks/archive/aeo19/.

EIA 2019b. "Henry Hub Natural Gas Spot Price." https://www.eia.gov/dnav/ng/hist/rngwhhdA.htm.

EIA 2019c. "Today in Energy: Daily Prices." https://www.eia.gov/todayinenergy/prices.php.

EIA. 2020a. *Annual Energy Outlook 2020 with Projections to 2050.* January 29. Washington, DC. https://www.eia.gov/outlooks/archive/aeo20/.

EIA. 2020b. "Table 7.2a Electricity Net Generation: Total (All Sectors)." *Monthly Energy Review.* September 24.

EIA. 2020c. "Electric Power Annual 2019." Washington, DC. October. https://www.eia.gov/electricity/annual/pdf/epa.pdf.

EIA. 2020d. "Table 1.7 Primary Energy Consumption, Energy Expenditures, and Carbon Dioxide Emissions Indicators." *Monthly Energy Review.* August 26.

EIA. 2020e. "Recent Data." Consumption and Efficiency. http://www.eia.gov/consumption.

Elliott, N., M. Molina, and D. Trombley. 2012. *A Defining Framework for Intelligent Efficiency.* Washington, DC: American Council for an Energy-Efficient Economy.

Energy Innovations. 2020. "Net-Zero Emissions Scenario." Policy Solutions. https://us.energypolicy.solutions/scenarios/home.

EPA (Environmental Protection Agency). 2019. *Global Non-CO_2 Greenhouse Gas Emission Projections & Mitigation 2015–2050.* Washington, DC.

EPA. 2020. *Inventory of U.S. Greenhouse Gas Emissions and Sinks: 1990–2018.* Washington, DC.

EPA Office of Air and Radiation, Climate Protection Partnerships Division (EPA). 2017. *National Awareness of ENERGY STAR for 2016: Analysis of 2016 CEE Household Survey.* Washington, DC.

EPRI (Electric Power Research Institute). 2019. *Renewable Ammonia Generation, Transport, and Utilization in the Transportation Sector.* Technology Insights Brief. Washington, DC.

ETC (Energy Transitions Commission). 2018. "Mission Possible: Reaching Net-Zero Carbon Emissions from Harder-to-Abate Sectors by Mid-Century." https://www.energy-transitions.org/wp-content/uploads/2020/08/ETC_Mission-Possible_FullReport.pdf.

Farbes, J., G. Kwok, and R. Jones. 2020. "Low-Carbon Transition Strategies for the Midwest." https://irp-cdn.multiscreensite.com/be6d1d56/files/uploaded/SDSN_Midwest%20Low%20Carbon%20Strategies_FINAL.20200713.pdf.

Friedmann, J.S., Z. Fan, and K. Tang. 2019. *Low-Carbon Heat Solutions for Heavy Industry: Sources, Options, and Costs Today.* New York, NY: Columbia University Center on Global Energy Policy.

Global Cement. 2019. "Cross River Partners with Svante for Carbon Capture and Storage." https://www.globalcement.com/news/item/10150-cross-river-partners-with-svante-for-carbon-capture-and-storage.

Global Cement. 2020. "Holcim US Invests in CCS Study at Portland Cement Plant." https://www.globalcement.com/news/item/10283-holcim-us-invests-in-ccs-study-at-portland-cement-plant.

Griffith, S., and S. Calisch. 2020. "Mobilizing for a Zero Carbon America: Jobs, Jobs, Jobs, and More Jobs." Rewiring America. https://www.rewiringamerica.org/jobs-report.

Haley, B., R. Jones, G. Kwok, J. Hargreaves, J. Farbes, and J. Williams. 2019. *350 PPM Pathways for the United States.* U.S. Deep Decarbonization Pathways Project. Evolved Energy Research. https://www.evolved.energy/post/2019/05/08/350-ppm-pathways-for-the-united-states.

Hand, M.M., S. Baldwin, E. DeMeo, J.M. Reilly, T. Mai, D. Arent, G. Porro, M. Meshek, and D. Sandor. 2012. *Renewable Electricity Futures Study.* Golden, CO: National Renewable Energy Laboratory.

Harper, G., R. Sommerville, E. Kendrick, L. Driscoll, P. Slater, R. Stolkin, A. Walton, et al. 2019. Recycling lithium-ion batteries from electric vehicles. *Nature* 575 (7781): 75–86.

Hsiang, S., R. Kopp, A. Jina, J. Rising, M. Delgado, S. Mohan, D.J. Rasmussen, et al. 2017. Estimating economic damage from climate change in the United States. *Science* 356(6345): 1362.

IEA (International Energy Agency). 2020a. *Energy Technology Perspectives 2020*. Paris, France.

IEA. 2020b. *Global EV Outlook 2020*. Paris, France.

ICCT (International Council on Clean Transportation). *The International Maritime Organization's Initial Greenhouse Gas Strategy*. Washington, DC.

Investing.com. 2019. "Gasoline RBOB Futures Historical Data." https://www.investing.com/commodities/gasoline-rbob-historical-data.

IPCC (Intergovernmental Panel on Climate Change). 2014. *Climate Change 2014: Synthesis Report*. Geneva, Switzerland.

IPCC. 2018. *Global Warming of 1.5°C*. Geneva, Switzerland: Intergovernmental Panel on Climate Change.

IRENA (International Renewable Energy Agency). 2018. "Hydrogen From Renewable Power: Technology Outlook for the Energy Transition." Abu Dhabi.

Jacobson, M.Z., M.A. Delucchi, G. Bazouin, Z.A.F. Bauer, C.C. Heavey, E. Fisher, S.B. Morris, et al. 2015. 100% clean and renewable wind, water, and sunlight (WWS) all-sector energy roadmaps for the 50 United States. *Energy & Environmental Science* 8(7): 2093–2117.

Jacoby, M. 2020. Tailor-made MOF readily catches and releases CO_2. *Chemical and Engineering News* 98(29).

Jenkins, J.D., M. Luke, and S. Thernstrom. 2018. Getting to zero carbon emissions in the electric power sector. *Joule* 2(12): 2498–2510.

Kelemen, P., S.M. Benson, H. Pilorgé, P. Psarras, and J. Wilcox. 2019. An overview of the status and challenges of CO_2 storage in minerals and geological formations. *Frontiers in Climate* 1: 9.

Kim, S.J., C.R. Tang, G. Singh, L.M. Housel, S. Yang, K.J. Takeuchi, A.C. Marschilok, et al. 2020. New insights into the reaction mechanism of sodium vanadate for an aqueous Zn ion battery. *Chemistry of Materials* 32(5): 2053–2060.

Krey, V., G. Luderer, L. Clarke, and E. Kriegler. 2014. Getting from here to there—Energy technology transformation pathways in the EMF27 scenarios. *Climatic Change* 123(3): 369–382.

Kriegler, E., J.P. Weyant, G.J. Blanford, V. Krey, L. Clarke, J. Edmonds, A. Fawcett, et al. 2014. The role of technology for achieving climate policy objectives: Overview of the EMF 27 study on global technology and climate policy strategies. *Climatic Change* 123(3): 353–367.

Kwok, G., J. Farbes, and R. Jones. 2020. *Low-Carbon Transition Strategies for the Southeast*. U.S. Deep Decarbonization Pathways Project. San Francisco, CA: Evolved Energy Research.

Lah, O. 2017. Decarbonizing the transportation sector: Policy options, synergies, and institutions to deliver on a low-carbon stabilization pathway. *Wiley Interdisciplinary Reviews: Energy and Environment* 6(6).

Larson, J., W. Herndon, M. Grant, and P. Marsters. 2019. *Capturing Leadership: Policies for the US to Advance Direct Air Capture Technology*. New York: Rhodium Group.

Larson, E., C. Greig, J. Jenkins, E. Mayfield, A. Pascale, C. Zhang, S. Pacala, et al. 2020. *Net-Zero America by 2050: Potential Pathways, Deployments, and Impacts*. Princeton, NJ: Princeton University.

Lazard. 2019. "Lazard's Levelized Cost of Energy Analysis—Version 13.0." https://www.lazard.com/media/451086/lazards-levelized-cost-of-energy-version-130-vf.pdf.

Lelieveld, J., and T. Münzel. 2019. Air pollution, chronic smoking, and mortality. *European Heart Journal* 40(38), 3204–3204.

Leuenberger, M., and R. Frischknecht. 2010. *Life Cycle Assessment of Battery Electric Vehicles and Concept Cars*. Schaffhausen, Switzerland: ESU-Services Ltd.

Li, J., E. Murphy, J. Winnick, and P.A. Kohl. 2001. Studies on the cycle life of commercial lithium ion batteries during rapid charge-discharge cycling. *Journal of Power Sources* 102: 294–301.

Liaukus, C. 2014. *Energy Efficiency Measures to Incorporate into Remodeling Projects*. Golden, CO: National Renewable Energy Laboratory.

Lutsey, N., and M. Nicholas. 2019. "Update on Electric Vehicle Costs in the United States through 2030." Working Paper. San Francisco, CA: International Council on Clean Transportation. April 2.

Mahajan, M. 2019a. "How To Reach U.S. Net Zero Emissions By 2050: Decarbonizing Buildings." Forbes. https://www.forbes.com/sites/energyinnovation/2019/11/05/reaching-us-net-zero-emissions-by-2050-decarbonizing-buildings/.

Mahajan, M. 2019b. "How To Reach U.S. Net Zero Emissions By 2050: Decarbonizing Electricity." Forbes. https://www.forbes.com/sites/energyinnovation/2019/11/12/how-to-reach-us-net-zero-emissions-by-2050-decarbonizing-electricity/?sh=4d9e62e749e7.

Majeau-Bettez, G., T.R. Hawkins, and A.H. Strømman. 2011. Life cycle environmental assessment of lithium-ion and nickel metal hydride batteries for plug-in hybrid and battery electric vehicles. *Environmental Science and Technology* 45(10): 4548–4554.

Marchau, V.A.W.J., W.E. Walker, P.J.T.M. Bloemen, and S.W. Popper. 2019. *Decision Making Under Deep Uncertainty: From Theory to Practice*. 1st ed. Switzerland: Springer International Publishing.

Mathy, S., P. Criqui, K. Knoop, M. Fischedick, and S. Samadi. 2016. Uncertainty management and the dynamic adjustment of deep decarbonization pathways. *Climate Policy* 16(1): 47–62.

McKinsey and Company. 2013. "Pathways to a Low Carbon Economy: Version 2 of the Greenhouse Gas Abatement Cost Curve." https://www.mckinsey.com/business-functions/sustainability/our-insights/pathways-to-a-low-carbon-economy.

McQueen, N., P. Psarras, H. Pilorgé, S. Liguori, J. He, M. Yuan, C.M. Woodall, et al. 2020. Cost analysis of direct air capture and sequestration coupled to low-carbon thermal energy in the United States. *Environmental Science and Technology* 54(12): 7542–7551.

Morrison, G.M., S. Yeh, A.R. Eggert, C. Yang, J.H. Nelson, J.B. Greenblatt, R. Isaac, et al. 2015. Comparison of low-carbon pathways for California. *Climatic Change* 13(4): 545–557.

NAM (National Association of Manufacturers). 2018. "2019 United States Manufacturing Facts." https://www.nam.org/state-manufacturing-data/2019-united-states-manufacturing-facts/.

NASEM (National Academies of Sciences, Engineering, and Medicine). 2019. *Negative Emissions Technologies and Reliable Sequestration: A Research Agenda*. Washington, DC: The National Academies Press.

Nemet, G.F. 2019. *How Solar Energy Became Cheap: A Model for Low-Carbon Innovation*. New York, NY: Routledge, Taylor and Francis Group.

OMB (Office of Management and Budget). 2020. "Historical Tables." Washington DC: The White House. https://www.whitehouse.gov/omb/historical-tables/.

Palacín, M.R., and A. de Guibert. 2016. Why do batteries fail? *Science* 351(6273): 1253292.

Park, J., S. Lim, and S. Yoo. 2019. Does combined heat and power mitigate CO_2 emissions? A cross-country analysis. *Environmental Science and Pollution Research* 26(11): 11503–11507.

Phadke, A., U. Paliwal, N. Abhyankar, T. McNair, B. Paulos, D. Wooley, and R. O'Connell. 2020. *2035 The Report: Plummeting Solar, Wind, and Battery Costs Can Accelerate Our Clean Electricity Future*. Berkeley, CA: Goldman School of Public Policy, University of California, Berkeley. September.

Pilorgé, H., N. McQueen, D. Maynard, P. Psarras, J. He, T. Rufael, and J. Wilcox. 2020. Cost analysis of carbon capture and sequestration of process emissions from the U.S. industrial sector. *Environmental Science and Technology* 54(12): 7524–7532.

Podesta, J., C. Goldfuss, T. Higgins, B. Bhattacharyya, A. Yu, and K. Costa. 2019. *A 100 Percent Clean Future*. Washington, DC: Center for American Progress.

Poyraz, A.S., J. Huang, S. Cheng, D.C. Bock, L. Wu, Y. Zhu, A.C. Marschilok, et al. 2016. Effective recycling of manganese oxide cathodes for lithium based batteries. *Green Chemistry* 18(11): 3414–3421.

Prehoda, E.W., and J.M. Pearce. 2017. Potential lives saved by replacing coal with solar photovoltaic electricity production in the U.S. *Renewable and Sustainable Energy Reviews* 80: 710–715.

PWC (PricewaterhouseCooper). 2017. "The World in 2050." https://www.pwc.com/gx/en/issues/economy/the-world-in-2050.html.

Psarras, P., J. He, H. Pilorgé, N. McQueen, A. Jensen-Fellows, K. Kian, and J. Wilcox. 2020. Cost analysis of carbon capture and sequestration from U.S. natural gas-fired power plants. *Environmental Science and Technology* 54(10): 6272–6280.

Reddy, S., and C. Freeman. 2018. "Fluor Solvent Evaluation and Testing (Project 70814)." Pacific Northwest National Laboratory. https://netl.doe.gov/sites/default/files/netl-file/C-Freeman-PNNL-Advanced-Solvent-Testing.pdf.

Reed, L., M. Dworkin, P. Vaishnav, and M.G. Morgan. 2020. Expanding transmission capacity: Examples of regulatory paths for five alternative strategies. *The Electricity Journal* 33(6): 106770.

Rightor, E., A. Whitlock, and R.N. Elliott. 2020. *Beneficial Electrification in Industry*. Washington DC: American Council for an Energy-Efficient Economy.

Rissman, J. 2019. "How To Reach U.S. Net Zero Emissions By 2050: Decarbonizing Industry." Forbes. https://www.forbes.com/sites/energyinnovation/2019/11/04/how-to-reach-us-net-zero-emissions-by-2050-decarbonizing-industry/.

Rissman, J., C. Bataille, E. Masanet, N. Aden, W.R. Morrow, N. Zhou, N. Elliott, et al. 2020. Technologies and policies to decarbonize global industry: Review and assessment of mitigation drivers through 2070. *Applied Energy* 266: 114848.

Schnitkey, G. 2018. "Weekly Farm Economics: Fertilizer Prices Higher for 2019 Crop." University of Illinois, Urbana-Champaign. *Farmdoc Daily* 8: 178.

SDSN (Sustainable Development Solutions Network). *Zero Carbon Action Plan*. New York, NY: Sustainable Development Solutions Network.

SEIA (Solar Energy Industries Association). 2020. *U.S. Solar Market Insight*. SEIA and Wood Mackenzie Power and Renewables. Washington, DC.

Sepulveda, N.A., J.D. Jenkins, F.J. de Sisternes, and R.K. Lester. 2018. The role of firm low-carbon electricity resources in deep decarbonization of power generation. *Joule* 2(11): 2403–2420.

Sepulveda, N.A., J.D. Jenkins, A. Edington, D. Mallapragada, and R.K. Lester. 2021. Evaluating the technology design space for long-duration energy storage and role in deep decarbonization of power systems. *Nature Energy* (in press).

Shonder, J. 2014. *Energy Savings from GSA's National Deep Energy Retrofit Program*. Oak Ridge, TN: Oak Ridge National Laboratory.

Steinberg, D., D. Bielen, J. Eichman, K. Eurek, J. Logan, T. Mai, C. McMillan, et al. 2017. *Electrification and Decarbonization: Exploring U.S. Energy Use and Greenhouse Gas Emissions in Scenarios with Widespread Electrification and Power Sector Decarbonization*. Golden, CO: National Renewable Energy Laboratory.

Sullivan, T. 2019. "2018 Building Energy Benchmarking." City of Seattle Open Data Portal. https://data.seattle.gov/dataset/2018-Building-Energy-Benchmarking/7rac-kyay.

Tallman, K.R., B. Zhang, L. Wang, S. Yan, K. Thompson, X. Tong, J. Thieme, et al. 2019. Anode overpotential control via interfacial modification: Inhibition of lithium plating on graphite anodes. *ACS Applied Materials and Interfaces* 11(50): 46864–46874.

Trembath, A., J. Jenkins, T. Nordhaus, and M. Shellenberger. 2012. "Where the Shale Gas Revolution Came From." The Breakthrough Institute. https://thebreakthrough.org/issues/energy/where-the-shale-gas-revolution-came-from.

Ungar, L., and S. Nadel. 2019. *Halfway There: Energy Efficiency Can Cut Energy Use and Greenhouse Gas Emissions in Half by 2050*. Washington, DC: American Council for an Energy-Efficient Economy.

United Nations. 2016. Amendment to the Montreal Protocol on Substances that Deplete the Ozone Layer, C.N.872.2016.TREATIES-XXVII.2.f § (2016). https://treaties.un.org/doc/Publication/CN/2016/CN.872.2016-Eng.pdf.

USG (U.S. Government). 2015. *United States Aviation Greenhouse Gas Emissions Reduction Plan*. Submitted to the International Civil Aviation Organization. https://www.icao.int/environmental-protection/Pages/ClimateChange_Action-Plan.aspx.

UT-Austin (University of Texas at Austin Energy Institute). 2020. "Levelized Cost of Energy Map, Version 1.4.0." https://calculators.energy.utexas.edu/lcoe_map/#/county/tech.

Vibrant Clean Energy. 2019. *Colorado Electrification and Decarbonization Study*. Boulder, CO: Community Energy, Inc.

Waisman, H., C. Bataille, H. Winkler, F. Jotzo, P. Shukla, M. Colombier, D. Buira, et al. 2019. A pathway design framework for national low greenhouse gas emission development strategies. *Nature Climate Change* 9(4): 261–268.

Wang, L., Q. Wu, A. Abraham, P.J. West, L.M. Housel, G. Singh, N. Sadique, et al. 2019. Silver-containing α-MnO_2 nanorods: Electrochemistry in rechargeable aqueous Zn-MnO_2 batteries. *Journal of the Electrochemical Society* 166(15): 3575–3584.

Webster, M., J. Arehart, R. Chepuri, J. D'Aloisio, K. Gregorian, M. Gryniuk, J. Hogroian, et al. 2020. "Achieving Net Zero Embodied Carbon in Structural Materials by 2050." White paper. Structural Engineering Institute's Sustainability Committee Carbon Working Group. American Society of Civil Engineers. https://www.asce.org.

White House. 2016. *United States Mid-Century Strategy for Deep Decarbonization*. Washington, DC.

Wilcox, J. 2012. *Carbon Capture*. New York, NY: Springer New York.

Wilcox, J. 2020a. The giving Earth. *The Bridge* 50(1): 43-50.

Wilcox, J. 2020b. "The Essential Role of Negative Emissions in Getting to Carbon Neutral." *Policy Digest*. Kleinman Center for Energy Policy, University of Pennsylvania. https://kleinmanenergy.upenn.edu.

Williams, J.H., B. Haley, F. Kahrl, J. Moore, A.D. Jones, M.S. Torn, and H. McJeon. 2014. *Pathways to Deep Decarbonization in the United States*. San Francisco CA: Energy and Environmental Economics, Inc.

Williams, J., R. Jones, G. Kwok, and B. Haley. 2018. *Deep Decarbonization in the Northeast United States and Expanded Coordination with Hydro-Québec*. Evolved Energy Research. https://www.evolved.energy.

Woo, J., and C.L. Magee. 2020. Forecasting the value of battery electric vehicles compared to internal combustion engine vehicles: The influence of driving range and battery technology. *International Journal of Energy Research* 44(8): 6483–6501.

Wood, D.L., M. Wood, J. Li, Z. Du, R.E. Ruther, K.A. Hays, N. Muralidharan et al. 2020. Perspectives on the relationship between materials chemistry and roll-to-roll electrode manufacturing for high-energy lithium-ion batteries. *Energy Storage Materials* 29: 254–265.

Wright, G.S., and K. Klingenberg. 2018. *Climate-Specific Passive Building Standards*. Golden, CO: National Renewable Energy Laboratory.

Zheng, J., Q. Zhao, T. Tang, J. Yin, C.D. Quilty, G.D. Renderos, X. Liu, et al. 2019. Reversible epitaxial electrodeposition of metals in battery anodes. *Science* 366(6465): 645–648.

CHAPTER THREE

To What End: Societal Goals for Deep Decarbonization

INTRODUCTION

Chapter 2 described the technological changes required to replace the current fossil fuel-based U.S. energy system with a net-zero carbon energy system. In Chapter 3, four priority social and economic overarching goals are identified to guide and evaluate those changes (Box 3.1). These strategies significantly expand how energy technologies and policies are typically assessed, modeled, and optimized, going beyond technical performance, cost, and reliability. They require more diverse voices and perspectives to be included in energy decision-making and new metrics for evaluating outcomes.

This expansion of the principles against which to measure U.S. energy transitions responds to three broad challenges: (1) the responsibility to ensure that the transition to a carbon-neutral economy benefits all Americans and addresses the harms that it creates; (2) the importance of establishing strong public support for action to decarbonize the economy; and (3) the possibility of leveraging opportunities created by the transition to advance a wide range of U.S. national priorities.

Replacing the systems that provide the United States with carbon-based fuels with carbon-neutral alternatives will require, in a very real sense, a fundamental transition of the U.S. economy. This transition has the potential to bring significant benefits to American families, workers, and businesses that go well beyond addressing climate change. As many parts of the country experienced during the early days of the COVID-19 crisis, for example, reductions in the use of carbon-based fuels will bring significant improvements in air pollution and public health, owing to lower rates of asthma, cardiovascular disease, and other pollution-linked diseases and mortality (Haines, 2017; Thakrar et al., 2020). The switch to clean energy can also advance U.S. economic leadership and competitiveness in global markets (Ladislaw and Barnet, 2019).

Further, the net-zero energy transition can reduce inequities in energy and transportation options that adversely impact significant numbers of Black, Indigenous, and people of color (BIPOC) (Drehobl et al., 2020; Bednar and Reames, 2020; Fleming, 2018), as well as communities experiencing economic decline and environmental injustice and

> **BOX 3.1**
> **SOCIAL AND ECONOMIC CRITERIA FOR THE TRANSITION TO A CARBON-NEUTRAL ECONOMY**
>
> - **Strengthen the U.S. economy** by accelerating innovation, advancing U.S. competitiveness, and creating high-quality jobs.
> - **Promote equity and inclusion** by ensuring a just and equitable distribution of costs, risks, and benefits and the effective participation of marginalized groups in transition decision making.
> - **Support communities, businesses, and workers directly impacted by the transition** to diversify economic development and financial resources and secure meaningful employment and environmental and health justice.
> - **Maximize cost-effectiveness** through the use of multicriteria methods and negotiations among politicians, stakeholders, and the public that maximize the ability to simultaneously achieve climate and societal goals at the lowest possible cost and with the optimum trade-off of co-benefits and risks. The committee recognizes that these other goals—regarding the economy, equity, and transition—are not as precisely defined and place constraints on cost-effectiveness.

other low-income and disadvantaged communities (Jessel et al., 2019; Colon, 2016; Shonkoff et al., 2011). However, the committee recognizes that environmental justice communities have concerns and want to ensure that emissions reductions happen in a way that all share in the benefits, such as is required under New York's Climate Leadership and Community Protection Act (S6599).

This transition may also provide employment opportunities. For example, one study estimates that cross-sector energy efficiency investments could add up to 660,000 more people working for a year (job-year) through 2023, 1.3 million added job-years over the lifetime of the investments, 910 million tons of reduced CO_2 emissions, and $120 billion in energy bill savings (Ungar et al., 2020).

At the same time, the United States cannot afford to ignore the difficulty and complexity of navigating one of the most disruptive economic transformations in U.S. history (Smil, 2010; Miller et al., 2013). Even at low levels of adoption, renewable energy is already transforming how electric utilities produce and sell energy, including shifts in business models, markets, prices, regulations, and the location of power production (Blackburn et al., 2014; Burger and Luke, 2017; Funkhouser et al., 2015). Some of the world's largest technology, finance, energy, and transportation firms have already initiated major changes to their operations that reduce the use of fossil fuels and increase energy efficiency, including Apple, Google, BP, General Motors, Ford, Delta Airlines, and BlackRock (Somini and Penney, 2020). These developments foreshadow the widespread changes that the transition to low-carbon technologies will bring to all sectors of the economy.

The rapid downsizing of carbon-intensive industries, the rise of low-carbon alternatives, shifts in energy geographies, and the reconfiguration and reorganization of electricity markets will also bring unprecedented change to significant factions of the U.S. workforce (NASEO and EFI, 2020). Throughout the economy, businesses and workers will need to adjust daily routines and work practices to the requirements of low-carbon electricity, energy, and transportation systems. Families and households will also experience significant changes, such as the need to replace gasoline-powered cars and trucks with electric vehicles (McCollum et al., 2018). Far more wide-ranging changes in household energy economics, practices, and behavior may also arise (Dietz et al., 2009)—for example, from distributed energy generation and storage technologies or programs for flexible demand through the active monitoring and regulation of household energy consumption. As these transformations proceed, they will intersect in predictable and unpredictable ways with other important changes in the U.S. economy, such as the damage wrought by COVID-19 (Henry et al., 2020; Sovacool et al., 2020), shifts in global trade (Byrne and Mun, 2003), and the rapid growth of automation, machine learning, and smart systems (Victor, 2019).

Meeting these challenges will require actions on the part of federal, state, and local governments, as well as businesses, workers, other institutions, families, and individuals. More intentional coordination will help the United States and its states, regions, and localities to navigate the complexities and uncertainties of economic transformation at the scales and on the timetables required to successfully combat climate change, to ensure the transformation is inclusive and equitable, and to provide support to businesses, workers, and communities as they face the consequences of change. The energy system, in particular, has a special responsibility to take the lead in transforming the U.S. economic system to replace the burning of fossil fuels with alternative, carbon-neutral and low-pollution means of creating, transporting, and using energy.[1]

---

[1] In using the phrase "the energy system," the committee recognizes that energy is, in reality, a complex system-of-systems that encompasses a wide range of technologies and societal, market, and regulatory arrangements responsible for the production and distribution of diverse energy resources, including fuels and electricity, as well as the myriad systems in which energy is used (e.g., buildings, transportation, food, communication, water, manufacturing, and more) for diverse human purposes. As indicated by its role in the economic system, these systems are deeply interconnected in their existing forms, and decarbonization will transform all of them and, in many cases, reconfigure their relationships—for example, via the electrification of vehicles, heat pumps, and other technologies. The reference to these diverse arrangements as the energy system is meant to encourage a comprehensive, integrated approach to decarbonization.

No longer can the United States tolerate delay in making the economic and technological changes necessary to combat climate change.

A SOCIAL CONTRACT FOR DECARBONIZATION

Because the transition to a low-carbon economy is likely to be disruptive and create uneven distributions of benefits, costs, and risks, U.S. energy policy in the 2020s will need to establish and maintain a strong social contract for decarbonization (see Box 3.2; O'Brien et al., 2009). In the absence of broad support from U.S. families, workers, businesses, and communities, progress is unlikely to proceed at the pace and scale required to achieve a carbon-neutral economy by 2050.

Polls show that, across the political spectrum, a significant majority of Americans support urgent efforts to combat climate change and decarbonize the economy (Leiserowitz et al., 2018; Roberts, 2020; Tyson and Kennedy, 2020). That support is likely to be tested, however, as the United States navigates the complexities of the changes required and the disruptions they bring to people's lives and livelihoods.

Research has demonstrated a "social gap" between widespread general support for renewable energy technologies yet relatively slow uptake (Dwyer and Bidwell, 2019; Rai and Beck, 2015; Boudet, 2019). Public perception and opposition can be roadblocks to a carbon-neutral transition (Firestone et al., 2017, 2020), especially where public engagement is perfunctory, carried out too late in the process, and where key decisions have already been made. These cases often exacerbate conflict among groups and catalyze opposition to new technologies and infrastructures. The deliberate undermining of public support for climate action through misinformation and the ways that publics are encouraged or discouraged from participating in governance processes can also significantly shape social responses to new technologies (Giordono et al., 2018; Hall et al., 2013). This is particularly relevant in the energy system, where there is often a lack of fairness and unequal distributions of power and resources in decision-making processes (Pezzullo and Cox, 2017; National Research Council, 2008).

There is no silver bullet for sustaining widespread public support for the transition to a carbon-neutral economy. That support will come only from persistent and sustained efforts on the part of civic, policy, labor, and business institutions in the energy system and beyond. A more coordinated, national effort is needed to proactively engage diverse publics and stakeholders (Dwyer and Bidwell 2019; Ashworth et al., 2011); to meaningfully integrate the social and economic dimensions of transitions into energy analysis and policy (Miller et al., 2015); and to work collaboratively with communities (Wyborn et al., 2019) to create a strong clean energy economy that supports a robust

BOX 3.2
A SOCIAL CONTRACT FOR DEEP DECARBONIZATION

The committee defines a social contract for deep decarbonization as a broadly shared understanding among the energy industry; local, state, and federal governments; and U.S. families, businesses, workers, and communities to support efforts to advance a transition to a carbon-neutral U.S. economy so long as that transition meets societally determined criteria.

Such a contract cannot be assumed to exist at present, nor will it result from naïve programs that seek only to "educate" the public. It must be created and nurtured via active public engagement that raises awareness and strengthens knowledge and learning as it listens and responds to individuals' and communities' concerns and incorporates diverse values into energy decisions. The committee believes that a principal way to get action on addressing climate change is to make sure that doing so also addresses the countless ways in which the U.S. energy economy has left people out, left some communities bearing excessive burdens of pollution and related public health problems, and led to communities dependent on fossil-energy resource extraction with limited lifetimes. The committee finds that making progress on mitigating the effects of climate change depends on navigating the energy transition in socially responsible ways.

Key considerations for such a contract include:

- Accepting a joint responsibility on the part of business, government, and civil society for transforming the U.S. economy and energy systems to carbon neutrality with sufficient rapidity to reduce the likelihood of extreme weather and climate risks and protect the environment for future generations.
- Honoring the contributions of energy workers to the nation's economy, including those displaced by the adoption of new energy technologies.
- Acknowledging interdependence among diverse stakeholders, sectors, and regions.
- Identifying, anticipating, assessing, and making transparent the societal and economic implications of future energy system design and use under diverse pathways to decarbonization.
- Engaging diverse communities and stakeholders in inclusive decision-making processes that allow participants to give full voice to their hopes and concerns about the current state of energy systems and the economy, decarbonization, and the energy and economic futures it will help bring into being.
- Providing financial support and capacity building to disadvantaged communities to ensure that they are able to effectively participate in transition decision making and contribute to the transition.
- Distributing the costs, benefits, risks, opportunities, and burdens of decarbonization fairly and equitably and redressing harms caused by the transition and by injustices and inequities that stem from existing energy systems.
- Leveraging energy innovation to create an economy that works better for all Americans, and especially for BIPOC (Black, Indigenous, people of color), women, rural, and low-income families, workers, and communities that have traditionally received a smaller proportion of the benefits of new energy technologies and systems or disproportionately borne their risks and burdens.

U.S. workforce and distributes the costs, benefits, risks, opportunities, and burdens of decarbonization as fairly and equitably as possible.

Generating sustained public support requires a multipronged approach, including public engagement to discover and embed community preferences in decision-making and a concerted effort to communicate the necessities, costs, benefits, and remedies of policy actions (Steg et al., 2015). It also needs to facilitate inquiry and dialogue about what those policies might mean for specific communities and how to apply policies equitably and effectively in different contexts (Kimura and Kinchy 2019), while systematically dismantling misinformation to minimize confusion and polarization (Farrell et al., 2019). Technology and infrastructure needs (as discussed in Chapter 2) toward deep decarbonization goals necessarily involve heterogeneous costs and benefits across communities and regions in the United States.

Inevitably, public support for necessary policy actions (see Chapter 4) will vary across U.S. regions based on perceptions of costs and benefits (Howe et al., 2015). Importantly, such perceptions are mediated through cognitive ideologies (e.g., individualistic versus egalitarian; Leiserowitz et al., 2013) and values (e.g., egoistic versus altruistic, Steg et al., 2015), which are relatively stable. Generating long-term public support will entail understanding those values and incorporating them into implementation to design strategies that are sensitive and responsive to local and contextual factors (Haggerty et al., 2018, Steg et al., 2015). Relatedly, to be effective, implementation strategies should take an integrated approach, anticipating barriers and challenges that communities and individuals might face with particular technologies or behaviors and crafting solutions that not only address immediate costs and benefits but also pay attention to ongoing informational and maintenance needs.

Achieving these goals will be arduous, but critical, and can only be accomplished through a deep commitment to working with relevant networks of trusted organizations and institutions and genuinely engaging communities in decision making (Berkes, 2009). The importance of public engagement is even higher in the early phases of the transition in order to establish a foundation of longer-term trust, cooperation, and transparency, without which broader and deeper scale-up actions necessary beyond 2030 could be crippled.

At the same time, it will be extremely important to prevent misinformation from continuing to exacerbate confusion, mistrust, and already polarized worldviews of the future of the energy system, thereby weakening public support for necessary policy actions (Farrell, 2019). Two things in particular could go a long way in taming the dangers of misinformation. First, financial disclosure and transparency requirements should be expanded and tightened to preclude proliferation of misinformation under the veil of secrecy and intractable affiliations (Farrell et al., 2019). Second, creating new forms of social interaction that bridge disconnected information-sharing systems has the potential to enable the cross-flow of information and building of linkages across diverse communities and value systems (Lewandowsky et al., 2017), thus helping rebuild a more foundational basis of trust.

Evidence strongly shows that, especially during times of significant technological change, robust public engagement using these kinds of strategies can deliver significant benefits with respect to both designing technological futures that effectively meet the needs of the public and strengthening public support for processes of change (Narrasimhan et al., 2018), especially where such engagement facilitates a bidirectional dialogue that connects national policy making with local communities (Devine-Wright, 2011; Petrova, 2013). This is particularly true where technological changes have substantial impacts on matters that are meaningful to members of the public (e.g., siting of new energy facilities near neighborhoods, the kinds of cars or light bulbs that are available to buy, energy costs, or the availability of alternative transportation modes) and where public engagement is carried out upstream, significantly in advance of proposed technological changes, and in a manner that allows for public input to make meaningful contributions to technology design or adoption (Wilsdon and Willis, 2004; Wiersma and Devine-Wright, 2014). Well-designed public engagement, including younger populations, also has the potential to significantly improve public literacy and learning on matters of concern, as well as more inclusive and constructive public decisions (Tierney and Hibbard, 2002; Bice and Fischer, 2020; McLaren Loring, 2007).

In light of these findings, it will be important for the United States to invest in innovative approaches to strengthen public engagement and participation in the design and deliberation of decarbonization pathways. These should include high-profile regional public dialogues and listening sessions organized by clusters of federal agencies in collaboration with state/regional governments and industry participation to discuss decarbonization pathways and goals and open conversations about questions of justice and inequality confronting communities in the context of decarbonization. It will also be important to set standards and resources for public participation in decarbonization planning processes by requiring a role for representatives of disadvantaged populations—low-income and communities of color—in advisory boards and other influential bodies to enable them to participate in meaningful ways. Standards should also mandate best practices in social impact assessment (Vanclay, 2003; Esteves et al., 2012), many of which have been neglected as federal project review has tilted heavily to focus solely on environmental criteria (Burdge, 2002).

Over the past decade, an increasingly broad coalition of groups has advocated that a low-carbon transition must be a "just transition": redressing the harms caused by the transition to a carbon-neutral economy in ways that ensure viable and thriving futures for the individuals, families, and communities whose lives and livelihoods have been disrupted (see Box 3.3; Carley and Konisky, 2020; Henry et al., 2020; Newell and Mulvaney, 2013; Sovacool et al., 2020). Similar to other movements, such as Black Lives Matter, that have highlighted persistent forms of injustice and economic insecurity in the U.S. economy and society, calls for a just energy transition highlight the importance of building a social contract for decarbonization that recognizes the ways that pathways differentially affect communities and using the resulting insights to design policies that create better, fairer, and more equitable outcomes. To address these concerns, a number of cities and states have already taken the lead in developing new approaches for evaluating and assessing the social and economic dimensions of pathways to decarbonization (e.g., City of Providence, 2019; California Energy Commission, 2018), which supplement more traditional methods for assessing the cost, reliability, and carbon footprint of new energy technologies and systems.

Over the next three decades, as U.S. cities, states, and companies move toward a carbon-neutral economy, they will make myriad decisions about how to reshape U.S. energy systems. Deep decarbonization offers a rare opportunity to deploy large-scale innovation in the energy system to advance an array of key U.S. national goals and objectives. In the 20th century, the electrification of cities, industry, and rural communities and the creation of world-leading automobile, oil, and gas industries played key roles in transforming America into a global economic and military power. Today, as described below, if the United States can leverage and sustain existing widespread public support for climate action and mobilize it in favor of the coordinated set of policy actions described in Chapter 4, the country has a similar opportunity not only to help minimize impacts of climate change but also to leverage deep decarbonization to strengthen U.S. economic leadership, reduce inequalities, and create a fairer and more just society.

On the other hand, failure to appropriately envision, evaluate, and integrate the social and economic implications of decarbonization into decision-making about pathways—and the attendant failure to secure a robust social contract with all segments of the American public that can overcome persistent and diverse efforts to undermine public will—poses stark risks to both the timing and achievement of deep decarbonization goals. These risks include erosion of popular and political support for both decarbonization as a goal and for specific policies and pathways to achieving it, higher costs, increased entrenchment of social division and inequality, persistent legacy threats to public and environmental health, and lost opportunities for systemic innovation to enhance near-term and long-term U.S. competitiveness.

BOX 3.3
THE JUST TRANSITION MOVEMENT AND THE U.S. EXPERIENCE OF ECONOMIC TRANSFORMATION

The crucial importance of attending to the wider societal and economic dimensions of decarbonization is rooted in the American experience of past economic transitions and failures. While the United States has never deliberately undertaken a transformation of critical infrastructures and industries as deep and rapid as that envisioned by decarbonization, workers and communities in many parts of the United States have experienced past periods of economic transition.

Prominent examples in living memory include the decline of industry and manufacturing in Rust Belt cities of the Midwest; the hollowing out of U.S. farming communities and the small towns that served them associated with agricultural transformation in the 1970s and 1980s; boom-bust cycles in the oil industry in places like Pennsylvania, Texas, and North Dakota; and the current collapse of the coal industry in West Virginia, Kentucky, and Wyoming. At the same time, there is a growing recognition that the U.S. economy has resulted in greater poverty and lower educational opportunities and upward mobility for some communities, including BIPOC (Black, Indigenous, people of color) (Table 3.3.1; Drehobl et al., 2020). These communities continue to suffer high rates of economic disenfranchisement and, as a result, high rates of illness and death in the COVID-19 pandemic (Oppel et al., 2020; Van Slyke, 2020). Acknowledging the need for decarbonization, proposals (Table 3.3.2) have been put forward to ensure that the U.S. transition to a low-carbon economy is a just transition and is informed by the experiences of vulnerable communities, including with environmental justice.[1] In addition to these just transition proposal examples, several pieces of proposed legislation have been drafted in Congress and political party platforms and political candidates have included just transition recommendations as well. The committee's policy recommendations for addressing the just transition goals are discussed in Chapter 4.

continued

BOX 3.3 Continued

TABLE 3.3.1 Vulnerable Groups in the Context of an Energy Transition

Stakeholder	Concerns in a Just Energy Transition
Coal, oil, and gas workers; power-plant workers; and other participants in fossil fuel-dependent economic activities, including manufacturing, operations and maintenance, and service industry jobs, e.g., in automobile parts or repairs or gas stations.	• Job loss • Local businesses dependent on business from energy industry employees • Accessible, alternative job training • Other economic concerns, including risks of insolvent benefit funds • Psychosocial impacts of lost occupational identity (Carley et al., 2018a; Carley and Konisky, 2020; Rolston, 2014)
Residents in places impacted by fossil fuel and renewable energy supply chains, the siting of energy facilities, and/or the decommissioning of legacy fossil-dependent facilities, including fenceline communities	• Economic opportunity versus local cost • Racial injustice • Environmental justice • Health and well-being • Psychosocial impacts (Jacquet and Stedman, 2013) • Consultation fatigue • Unreclaimed infrastructure and associated health risks
Native American nations and rural communities whose economies, tax revenues, or lands are currently dependent on or impacted by coal and oil and gas development or potentially impacted by future renewable energy development	• Economic opportunity versus local cost • Racial injustice • Environmental justice • Health and well-being • Less tax revenue for schools and other publicly supported services
Clean energy industry workers and workers in the energy efficiency industry	• Looking for (better, long-term) jobs • Professional development/advanced training
Communities facing high energy costs and burdens that contribute to perpetuating or exacerbating poverty	• Affordable electricity • Accessibility and connectivity to immediate and distant areas/regions • Access to opportunities and financing to improve infrastructure to reduce costs and take advantage of renewable energy opportunities

BOX 3.3 Continued

TABLE 3.3.2 Just Transition Proposals and Proponents

Proposal	Proponents/Authors	Key Themes	Notable Recommendations
The National Economic Transition Platform (2020)	Just Transition Fund and coalition	Community-based, reclamation, infrastructure, bankruptcies, access to federal resources, workforce development, restorative economic development.	
Just Transition Platform (2020)	European Union (EU)	Development, reskilling, and environmental rehabilitation; social and economic effects of the transition, focusing on the regions, industries, and workers who will face the greatest challenges.	The Just Transition Platform aims to assist EU countries and regions to unlock the support available through the Just Transition Mechanism. This platform will provide a single access point for support and knowledge related to the just transition. Just Transition Mechanism (JTM) is a key tool to ensure that the transition toward a climate-neutral economy happens in a fair way, leaving no one behind. It provides targeted support to help mobilize at least €150 billion over the period 2021–2027 in the most affected regions, to alleviate the socioeconomic impact of the transition.

continued

BOX 3.3 Continued

TABLE 3.3.2 Continued

Proposal	Proponents/Authors	Key Themes	Notable Recommendations
Guidelines for a just transition toward environmentally sustainable economies and societies for all	International Labour Organization	Social consensus, workers' rights, gender equity, workforce support and development, no "one size fits all" approach, United Nations Sustainable Development Goals (UN SDGs)	Summarizes opportunities and challenges to a just transition. Developed seven guiding principles.
Guiding Principles and Lessons Learnt for a Just Energy Transition in the Global South (2017)	Friedrich-Ebert-Stiftung	Climate ambition, Nationally Determined Contributions-Sustainable Development Goal (NDC-SDG) alignment, decent work and vulnerability focus, social equity, gender equity, due participation, good governance, respect for human rights	Developed a set of eight just energy transition principles designed to make justice applicable to energy transition processes in developing countries, which go beyond an abstract call for justice, including climate, socioeconomic, and political dimensions in a balanced way to reflect the legitimate justice claims of a broad range of potential allies for a just energy transition alliance.

---

[1] The U.S. Environmental Protection Agency defines environmental justice (EPA, n.d.) as: "the fair treatment and meaningful involvement of all people regardless of race, color, national origin, or income, with respect to the development, implementation, and enforcement of environmental laws, regulations, and policies," including the "same degree of protection from environmental and health hazards, and equal access to the decision-making process to have a healthy environment in which to live, learn, and work."

LEVERAGING DEEP DECARBONIZATION FOR ECONOMIC AND SOCIAL INNOVATION

The committee recommends four social and economic criteria for evaluating pathways to a carbon-neutral economy and informing the decisions that will need to be made over the next several decades, to bring about a just transition. These four criteria are:

1. strengthening the U.S. economy;
2. promoting equity and inclusion;
3. supporting communities, businesses, and workers impacted by the energy transition; and
4. ensuring cost-effectiveness.

Each of these criteria reflects an important plank in the U.S. social contract for deep decarbonization because they address the critically important question: To what ends, beyond carbon-neutrality, should the United States pursue deep decarbonization?

These four considerations, described below, are not necessarily comprehensive. The transition to a carbon-neutral economy will bring both significant benefits and challenges for U.S. national security that require extensive analyses that are beyond the scope of this committee. Examples include the implications for the fueling and powering of U.S. defense systems and military operations, critical material and equipment supply chains, emergent vulnerabilities to disruption owing to climate change, and impacts of regional and global alliances. The transition to a carbon-neutral economy will significantly reduce U.S. health and environmental risks, especially in communities that have historically suffered from higher levels of air pollution owing to the combustion of carbon-based fuels. A full assessment of these benefits and considerations is also beyond the scope of this interim report.

As discussed further in Chapter 4, navigating the coming transition successfully will also require strengthening the capacity of energy regulatory and governance institutions to address the complex and interdependent choices these institutions will face in the coming decades and bolstering processes to strengthen the participation of diverse voices and put them on more equal footing with traditional energy stakeholders. Transition policies will require extensive and new forms of coordination across sectors (e.g., between electricity and transportation), across jurisdictions (e.g., between cities and suburbs and their rural neighbors), among utilities (e.g., within regional markets), and between the public and private sectors (e.g., between utilities and cities). These and other relevant considerations should also be part of any comprehensive approach to decarbonization policy and planning.

The committee recognizes that the U.S. Congress and President, state legislatures and governors, city councils and mayors, energy company and utility boards of directors and chief executive officers, civic and business leaders, tribal leaders, and ordinary Americans will bring diverse values and perspectives to choices about how to achieve a carbon-neutral economy. Their decisions will also be shaped by a variety of local and regional considerations such as the differential availability of low-carbon energy resources such as wind and sunlight, the needs of local and regional economies, the configuration of local and regional transportation systems, and the local and regional legacies of carbon-based energy. It is essential that this diversity of values, perspectives, needs, and contexts be given due weight and influence in transition planning and policy.

Approaches that weigh one of these criteria very heavily while neglecting the others are neither desirable nor are likely to be sustainable or secure public support over a multidecade period. This perspective renders inadmissible policy proposals that focus, for example, only on cost minimization and effectiveness while neglecting distributional effects, as well as instruments that singularly prioritize industrial competitiveness while disregarding cost-effectiveness or the needs of communities impacted by the transition. Pragmatic approaches to decarbonization will achieve balance across all four criteria detailed below.

Strengthen the U.S. Economy

The first criterion is that deep decarbonization pathways should strengthen the U.S. economy by accelerating innovation, advancing U.S. competitiveness in the global economy, and creating high-quality jobs, in relation to a clean energy future. Assessing success in creation of high-quality jobs will require development of the definition for "high-quality jobs" as discussed in Chapter 1. Ensuring that decarbonization advances the U.S. economy and benefits U.S. firms and workers will help maintain the social contract for deep decarbonization.

In the United States, the energy transition is expected to generate public and private investments in new energy technologies and infrastructure worth several trillion dollars (IRENA, 2019). Worldwide, total investment by 2050 in the energy system is estimated at $110 trillion (IRENA, 2019). A significant fraction of these investments is already committed, in the form of targets set by companies, countries, states, and cities. BlackRock has announced, for example, that it intends to put the low-carbon energy transition at the center of its $7 trillion investment portfolio (Coumarianos and Norton, 2020). The European Union has pledged to reduce net carbon emissions to zero by 2050, with an anticipated $1 trillion in public investments in clean energy in the next decade (Krukowska and Chrysoloras, 2019; Vetter, 2020). Numerous companies, cities, and states in the United States have made similar commitments, including several of the largest U.S. electric utilities (Porter and Hardin, 2020). Volkswagen has indicated its commitment to increase production of electric cars for the masses (Ewing, 2019).

These commitments and investments present a unique opportunity for American businesses and workers to participate in the creation of an entirely new industry and global infrastructure for clean energy comparable to the creation and growth of the oil, gas, and automobile industries over the past 150 years, including the potential to ensure that the benefits of the clean energy economy are equitably shared among all Americans. Missing this opportunity would create enormous economic headwinds for the United States deep into the 21st century.

Energy systems are deeply embedded in our economy and enable it to operate. It is a significant employer and a critical infrastructure that supports all economic activity. In fact, according to Energy Entrepreneurs (E2) latest report, the clean energy workforce in the United States reached 3.3 million jobs by the start of 2020, and it continues to grow for the fifth straight year (Energy Entrepreneurs, 2020). New energy infrastructure will require industrial products, manufactured goods, business services, and new jobs in construction and operations. Better coordination and planning will be required to ensure that U.S. deep decarbonization pathways recognize these linkages between energy and the U.S. economy and leverage them to promote U.S. leadership in the development and manufacturing of new energy technologies, to provide low-cost, reliable, and clean energy to U.S. businesses, and to grow significant new energy industries and associated high-quality jobs.

Clean Energy Contributions to U.S. Innovation, Competitiveness, and Jobs

The committee defines the objectives of leveraging investments in the energy transition to strengthen the U.S. economy in terms of four goals:

Goal 1: Deep decarbonization policy in the 2020s should lay the groundwork for ensuring that the United States has access to growing, reliable, low-cost, clean energy supplies as an essential foundation for a sustainable, resilient, diversified, equitable, and growing economy throughout the 21st century. A thriving, sustainable 21st century U.S. economy requires a secure and abundant supply of low-cost, clean energy. Achieving this goal will require significant investments in clean energy innovation, including strategies for development and widespread deployment of new clean energy technologies and significantly reducing their costs over time.

There are many options available and pathways to meet carbon neutrality in the U.S. economy by 2050, as well as significant variability in regional needs and contexts. A key facet of this goal is also to create flexibility in the options available to the United States for achieving deep decarbonization targets.

Goal 2: Clean energy transitions should accelerate and leverage U.S. strengths in innovation. The United States is a world leader in innovation. Key clean energy technologies have been invented and pioneered in the United States, and the United States leads the world in research investments in clean energy and in the development of a number of future technologies that are likely to play a significant role in achieving deep decarbonization. The United States currently struggles, however, to leverage its leadership in clean energy research and innovation into leadership in clean energy markets and supply chains. U.S. policy should find ways to ensure that the United States maintains its leadership in the discovery, invention, and development of innovative clean energy technologies, while also leveraging that innovation to ensure that the United States is positioned to manufacture and supply the technologies necessary to create a vibrant clean energy infrastructure as a basis for a thriving economy.

Goal 3: Clean energy transitions should enhance and leverage the global competitiveness of U.S. firms. Global markets for clean energy technologies and services are expanding rapidly and are expected to continue to do so for the next several decades at very high annual rates of growth. This growth presents a significant opportunity for U.S. companies, if the United States is able to establish globally competitive industries in key technology markets. U.S. policy should make sure that U.S. companies are positioned to successfully compete in global clean energy markets and do so in ways that are able to be sustained and resilient in the face of future global shocks.

Goal 4: Clean energy transitions should grow the U.S. workforce through the creation of new, high-skilled, high-wage jobs. The U.S. energy industry is a major employer, and this position of importance in the U.S. workforce will continue into the future as the United States revamps the energy system to meet deep decarbonization targets. The U.S. Bureau of Labor Statistics, in its Occupational Outlook Handbook, also notes that wind turbine service technicians and solar photovoltaic installers are forecasted to be the first and third, respectively, fastest growing occupations between 2019 and 2029 (U.S. Bureau of Labor Statistics, 2020). However, the transition from the existing U.S. energy workforce to the energy workforce of the future will pose significant challenges for U.S. energy workers, their families, and communities dependent on their incomes, and requires careful consideration of individual, household, and community transition planning. The United States is no stranger to the economic challenges posed by disruptive innovation, but going forward it must do significantly better at cushioning the impacts and maximizing economic benefits of rapid technological changes and anticipating and proactively addressing the transition needs, especially for industries most impacted, of communities and regions. The future of the U.S. energy workforce is critically dependent on U.S. leadership in the clean energy economy and on ensuring that the emerging clean energy economy supports high-quality jobs.

How Deep Decarbonization Innovation Strengthens the U.S. Economy

Beginning in World War II, the United States learned the importance of public and private investments into research, development, demonstration, and deployment of technologies as well as the necessity of having a well-educated workforce that could be deployed in factories and laboratories across the nation. Innovation is a crucial engine for technology discovery and development, as well as for long-term reductions in technology costs and improvements in quality. Innovation is also an important engine for entrepreneurship, especially in tech-heavy sectors and, thus, fundamentally linked to the potential for long-term job creation in the U.S. economy and the ability for the economy to successfully navigate disruptive technology transitions. Last, innovation is a necessary, albeit not sufficient, condition for U.S. competitiveness in a global economy in which innovation is now understood as the foundation for long-term economic security and in which all countries now invest heavily.

Decarbonization requires significant new innovation (Chu and Majumdar, 2012; IEA, 2020b). Many of the technologies necessary for the initial pursuit of deep decarbonization strategies are already established industries and several have already significantly fallen in costs (Wiser and Bolinger, 2019). The U.S. Department of Energy (DOE) Sunshot program, for example, helped bring about reductions in solar energy costs by 80 percent from 2010 to 2020, with the goal of further cost reductions of another 50 percent by 2030 (DOE-SETO, 2020). Further cost reductions in solar and wind would continue to accelerate adoption of low-carbon technologies and significantly reduce the overall costs of a transition. Future cost reductions are also essential in other core low-carbon technologies, for example, lithium-ion batteries, to achieve cost-effective decarbonization pathways. At the same time, there is not a one-size-fits-all decarbonization pathway, especially in the 2035–2050 time period. It is important to keep a wide array of options open, which will include significant needs for innovation in an array of potential low-carbon technology domains, for example, hydrogen, direct air capture of carbon dioxide, and vehicle-to-grid technologies (IEA, 2020a). Innovation in these domains will help ensure that the United States has the flexibility to respond quickly to rapid changes in energy markets, climate change impacts, and technological trajectories as it pursues deep decarbonization.

It is important to note both historical and current trends in federal investments in research and development (R&D) and how federal investments compare to other countries. Federal government R&D peaked in 1964 at 1.9 percent of GDP but has steadily declined since to just 0.6 percent of GDP in 2017. In that same time period, private R&D investments rose from 0.9 percent of GDP in 1964 to 1.95 percent of GDP in 2017 (Boroush, 2020; CRS, 2020). Working synergistically, public and private investments in R&D led to the birth and growth of large numbers of vibrant new U.S. industries that led the world in computers, data, information, pharmaceuticals, communication, nuclear energy, satellites, space exploration, GPS, solar, and aviation, among many others (Ruttan, 2006; Jenkins et al., 2010; Nemet, 2019; Gordon, 2016). The information revolution was led by the United States, for example, leading to the creation of whole new industries initially dominated by American firms and, still today, with major American firms at their apex.

Today, while the United States is the largest R&D investor globally in aggregate (soon to be surpassed by China if recent trends continue), it has fallen to 10th in terms of R&D intensity (R&D investments as a percentage of GDP). In terms of the average annual growth rate of domestic R&D expenditures, the United States ranks 6th, at 4.3 percent per year, compared with China at 17.3 percent per year and South Korea at 9.8 percent per year (Boroush, 2020). This relative decline in rates of new investment in R&D have created challenges for U.S. firms and the economy as a whole in maintaining their competitiveness in global markets. The U.S. first-mover advantage in many technologies, such as information technology and artificial intelligence, has since eroded owing to a strong challenge from China in particular (Allison, 2019), which has concentrated public investments in key technologies in order to secure long-term market advantages. Some economists argue that the incremental gains in productivity from the IT revolution are diminishing fast and will not sustain the United States as a major source of economic growth in the future, especially with rising economic inequality (Gordon, 2016).

The United States was a leader in developing clean energy technologies like wind and solar, but has ceded much of that leadership to other countries as these technologies have matured and become cost-competitive (Lewis, 2014; Platzer, 2012)—for example, in the solar industry, where Chinese firms today hold most of the leading positions. The lack of sustained policy signals to industry (Nemet et al., 2017), such as a national clean energy standard or a feed-in tariff, along with inconsistent incentives such as the intermittent production tax credit for wind technologies, have failed to create the markets necessary to support robust domestic manufacturing. This disturbing trend puts the United States at risk of losing out in the global clean energy industries. Alternatively, coherent, long-term policies to support the transition to a carbon-neutral economy can be leveraged to regain global leadership and competitiveness in clean energy technology, modernize and transform the U.S. manufacturing base, and create a new generation of clean energy jobs (Lester and Hart, 2012).

The United States is well positioned for economic growth in a low-carbon, resilient economy. The nation has a strong tradition of entrepreneurship and innovation, the two key ingredients for disruptive growth (e.g., Schumpeter, 1934). Owing to its strong commitment to public education, it has long had an educated and well-trained workforce. The nation has ample land, so should not face physical constraints on green energy infrastructure. It has abundant supplies of every type of major low-carbon energy resource, although these are differentially distributed across the nation. The United States can thus count on sufficient amounts of energy in a secure, carbon-neutral future and is poised to exploit renewable energy resources much more pervasively than it has in the past. This will be especially true if the United States can secure a significant share of domestic and global markets for the clean energy technologies that will be necessary to achieve a carbon-neutral U.S. energy infrastructure by 2050.

The renewable energy industry and energy efficiency industry are both high-growth sectors of the U.S. economy, and both are likely to continue to grow significantly under the scenarios laid out in Chapter 2 for transitioning the U.S. economy to carbon neutrality. Many trends in these industries are strong, in terms of the growth of high-quality jobs, but caution needs to be taken to guard against inappropriate treatment of workers, especially in the context of anti-competitive Chinese policies that have undermined profitability for decades in the renewable energy industry (Wu, 2019). Some criticisms and concerns that have been raised focus on the responsibility the clean energy industry has with respect to its workforce, including low wages for workers, lack of training and skills development programs, lack of access to career pathways, use of temporary workers without benefits, inadequate protection for health and safety, exploitative business models, and misclassification of workers (Mulvaney, 2014; Newell and Mulvaney, 2013). States such as California have worked to address this criticism by making a commitment to ensure that all state residents thrive in a carbon-neutral energy transition and by developing a framework to implement their ambitious plans (Roth, 2020; Zabin, 2020). Policies in Chapter 4 address these concerns regarding polices related to incentives for community benefits and good wages. However, recent research has demonstrated the ability of innovative programs to successfully integrate equity considerations into greenhouse gas reduction efforts and leverage tracking and feedback to ensure high-quality jobs are part and parcel of the transition, thus demonstrating that the "jobs versus environment" debate is a false choice and getting both is possible (Zabin, 2020). Ultimately, an important goal of the journey in the United States to a carbon-neutral economy should be to develop a comprehensive, integrated approach to a clean energy transition that ensures that the U.S. energy workforce is larger, better compensated, and more secure, overall, in 2050, than it is today.

Promote Equity and Inclusion

The second criterion to evaluate the design of clean energy transitions is that clean energy transitions should help to create future U.S. energy systems that are more just, equitable, and inclusive. This requires careful attention to ensure that both the processes through which decisions about energy transitions are made and the outcomes of clean energy transitions are more inclusive of the full array of voices of workers and communities with stakes in the future of U.S. energy and that these diverse communities are treated fairly and equitably.

Defining Equity and Inclusion for Clean Energy Transitions

The committee defines just, equitable, and inclusive transitions in terms of three key normative goals:

Goal 1: The benefits of clean energy should be distributed broadly and equitably, and likewise its burdens, risks, and costs. Clean energy systems will create a variety of benefits, including access to clean energy sources, opportunities for business and investment, cleaner environments, new jobs, and more. They will also create a variety of new costs, risks, and burdens associated with, for example, paying for the transition, the siting of new facilities and factories, payments for energy and energy services, purchase of new equipment (e.g., heat pumps or cars), decline in fossil-fuel industry jobs, and exposure to hazardous materials. Careful attention should be paid to who reaps these benefits and pays these costs and whether they are fairly and equitably distributed across groups and across the country. This will require advancing robust frameworks for assessing the equity implications of clean energy policies and development. Several federal and state policy frameworks already mandate analysis of equity dimensions of government decision making. Additionally, some local policy frameworks such as in the City of Minneapolis' Climate Action Plan, call for reporting to include monitoring progress annually, inclusive of equity indicators (City of Minneapolis, 2013). These range from considering environmental justice risks in permitting and environmental review (Ramos and Pires, 2013) to designing implementation of grant programs to prioritize access for disadvantaged groups (CPUC, 2019). Although not at the scale needed for net-zero policy, these programs provide important lessons for developing federal equity standards and rules.

Goal 2: The voices and perspectives of current and historically marginalized groups should be clearly and effectively included in and integrated into clean energy planning and decision making. Ensuring meaningful public participation by those most affected is not just a matter of ethics. It is critical to ensuring that policies are well designed to address equity and work for all Americans, as well as to upholding the U.S. commitment to democratic decision making that is open to and inclusive of all voices. Without attention to equity, the policies and implementation will not garner sustained public support and may face significant opposition or backlash. It has been seen in countries like France that policies that did not sufficiently address economic equity led to widespread protests (Williamson, 2018). Similarly, policies designed without appropriate input from diverse communities may fall short of long-term carbon neutrality goals. For example, California's Assembly Bill (AB 32) was less inclusive of environmental justice groups, and new companion legislation (AB 617) was designed to overcome the shortcomings and empower communities for addressing local environmental issues (Fowlie et al., 2020). Many important sources of carbon dioxide emissions in the U.S. economy are widely distributed across all communities, including buildings, equipment, and automobiles in the possession of households and businesses in low-income, indigenous, and rural communities and communities of color and people with disabilities, many of which will struggle to transition to carbon neutrality without policies that support and reflect their distinct needs and contexts.

Significant and sustained efforts will be required to strengthen and expand public participation in energy decision making in order to counter both misinformation and efforts to hamper public engagement in climate policy that threaten the social contract for deep decarbonization (Bush, 2019; Whitehouse, 2015). Policy and financial commitments will be needed to ensure not only that decision-making processes that shape the future of energy are transparent to and inclusive of the voices of diverse communities but also that these communities have the resources and are able to develop the capacity to participate effectively. This has been a significant emphasis in recent proposed federal law (e.g., the Environmental Justice for All and Climate Equity Acts). In developing appropriate policies to support enhanced participation in energy decision making, the United States should be guided by the experience of U.S. environmental and climate justice organizations who, despite being significantly underfunded (Taylor, 2014), have worked to represent many of these communities, win public participation rules that ensure that their communities have the resources and capacity to participate meaningfully in decision making, and strengthen public education and accountability. Scholars have recommended building on President Bill Clinton's Environmental Justice Order to incorporate climate and energy justice communities and organizations (White-Newsome, 2016). According to the National Economic Transition Platform (Just Transition Fund, 2020), a priority is to "build the capacity of local community-based leaders and organizations and facilitate community-driven planning processes and on-going program development and implementation. This is achievable through training and mentorship programs, grant funding to directly support salaries and materials needed for planning and program implementation, support from resource experts, and other technical assistance."

Partnerships with civil society organizations and philanthropic foundations have helped governments significantly strengthen public support for multi-billion-dollar investments toward the creation of a carbon neutral economy, improved health, and greater equality (see, e.g., State of California, 2020). Valuable insights can also be drawn from international experiences, such as the work of the United Nations to enshrine the principle of free, prior, and informed consent (UN FAO, 2016) as a key right of indigenous communities where decisions impact those communities or their lands (Dunlap, 2017; Mercer et al., 2020; Papillon and Rodon, 2017). Given the need to sustain a strong social contract for deep decarbonization, it is critical for policy makers and philanthropic actors to continue to work together to strengthen public participation and climate equity by scaling up support to organizations representing environmental justice communities (Lerza, 2011) and strengthen support for public participation in energy and economic transformation (Renn et al., 2020). Additionally, it is vital that philanthropic organizations prioritize addressing both the severe racial justice and equity disparities in their funding of climate non-governmental organizations (NGOs) (Baptista and Perovich, 2020), and the diversity of their board and staff advisors (Taylor, 2014), as the public sector-philanthropic partnerships become more prevalent as one tool to hold governments accountable for their contributions to equitable outcomes (Ferris and Williams, 2012). Best practices from successful public sector-philanthropy partners are needed to be replicated and scaled (Ferris and Williams, 2012), especially when it comes to equitable funding.

Goal 3: Clean energy transitions should reduce or eliminate economic inequalities and insecurities exacerbated by U.S. energy systems. All families and businesses consume and pay for energy, in some form. For most, energy bills are an ordinary cost of living and doing business in a modern society. Low- and moderate-income communities and businesses, however, often confront high financial burdens from energy costs that undermine economic security, force trade-offs between energy, food, and other basic necessities, recurrently threaten shutoffs of energy services that pose health risks during extreme heat and cold events, and create stresses that undermine productivity and well-being (Carley and Konisky, 2020; Finley-Brook and Holloman, 2016; Jessel et al., 2019; Madlener, 2020). Negative feedback can further reinforce the linkages between energy and poverty—for example, by limiting the ability of low-income communities to invest in energy efficiency improvements or higher quality products—thus perpetuating higher energy costs and so reducing the ability to pay for energy. These communities also bear a disproportionate burden of environmental and health risks associated with energy systems, across the life cycle from extraction, generation, and distribution to end-of-life and legacy infrastructure risks (Bullard, 2015; Bednar and Reames, 2020; Liévanos, 2018). None of this is necessary. Environmental justice mapping and screening tools and reporting exist that can be used to identify the communities most affected by sources of pollution and where people are often vulnerable to the effects of pollution. Currently environmental justice screening and mapping tools are outdated and are not sufficiently enforceable.

Furthermore, strategies for innovative clean energy transitions are positioned to reduce energy burdens and create solutions that are economically generative for these communities—for example, through opening up ownership, investment, and employment opportunities in clean energy to low- and moderate-income communities and enhancing the value of energy for low-income users. This will be particularly important as the country pursues decarbonization initiatives that extensively implicate infrastructure in low-income communities—for example, in improving energy efficiency and electrifying energy end uses in residential and commercial buildings, as well as electrifying vehicles. All of these are likely to impose significant costs on low-income communities (or to risk failing to achieve decarbonization goals), unless explicit attention is paid in policy design to this challenge (Miller et al., 2015). A number of cities and states have developed innovative policies for directing new revenues from decarbonization investments into projects to benefit low-income communities. Examples include the following:

- The state of California established an economy-wide cap-and-trade program to reduce greenhouse gas (GHG) emissions, which provides revenues to the state from the sale of GHG emission allowances. A significant portion of the proceeds from these auctions are invested in underserved communities. In 2019, for example, these revenues provided more than $1 billion for new projects implemented in disadvantaged communities and low-income communities and households. Cumulatively, $5.3 billion in projects have been implemented since the start of the program, with 57 percent of those investments benefiting priority populations. The funds have been used to "expand low-carbon transportation options, place affordable housing adjacent to transit and job centers, decrease the risk of catastrophic wildfires, and improve water-use efficiency," as well as research, planning support, work training, and technical assistance to local community groups (State of California, 2020). It should be noted that while funds have been distributed to priority populations, the extent to which improvements have been made is unclear, and criteria pollutant hot spots may still be present.

- In 2018, voters in the City of Portland, Oregon, approved a ballot measure to establish the City's Clean Energy Community Benefits Fund. Through this program, large companies contribute 1 percent of their revenues to a locally managed fund to support local clean energy, energy efficiency, and climate-justice projects. The program has been anticipated to generate $44 million to $61 million a year for grants to support jobs in clean energy sectors for underserved and energy-burdened communities in the Portland area. The fund is guided by a diverse advisory board that includes the communities bearing the greatest burden, as well as business and policy leaders (City of Portland, 2020).
- The Philadelphia Energy Authority (PEA) was created in 2010 to address energy affordability and sustainability issues. The PEA views energy as a tool for impact and promotes economic development, creates jobs, alleviates poverty, and supports efforts to improve public health. The Philadelphia Energy Campaign, which was launched in 2016, includes an investment in energy efficiency and clean energy projects of $1 billion over 10 years and focuses on municipal buildings, K-12 schools, affordable housing, and small businesses. The campaign, through 2019, has seen some early successes including $136 million in active projects and 1,301 new jobs. Other important outcomes of PEA's initiative include plans for the country's largest solar project, which will cover about 22 percent of the city government's electricity use, and a multifamily affordable housing pilot project aimed at generating 15 to 30 percent energy savings for renters and supports building owners investing in energy efficiency and smart grid technologies. Another new PEA program is the Solarize Philly program, which received a DOE Bright Solar Futures award and is the country's first program that provides vocational training for high school students to become solar installers. Through 2020, the large program involves a total of 700 solar contracts, a total capacity of 3 MW, an investment of $11 million in efficient and clean energy, added 52 new jobs, and has 6,500 households signed up. These efforts contribute to meeting the city's climate commitments of 100 percent renewable electricity by 2030 and reducing carbon emissions by 80 percent by 2050 (NASEM, 2019).

Rationale for Just, Equitable, and Inclusive Energy Futures

The rationale for ensuring that clean energy futures are just, equitable, and inclusive is grounded in a set of philosophical, pragmatic, and aspirational commitments.

Philosophical foundations of just transitions: The energy system is implicated in a range of historical, present, and potential future forms of injustice and inequality that should be redressed, anticipated, and proactively avoided in socially responsible transitions to carbon-neutral futures. All too frequently, local communities fail to derive meaningful benefits from energy infrastructures built in or near them, and they can experience significant negative health or environmental impacts (Bridge et al., 2018; Dao, 2020). These challenges are experienced throughout energy supply chains, from resource extraction to refineries and pipelines to power plants and transmission lines.

Energy systems as currently constituted often create financial, psychological, and other burdens on low-income communities that exacerbate poverty, inequality, and economic insecurity via a wide variety of mechanisms. Low-income communities also often suffer from lower quality energy infrastructures—for example, less reliable, less efficient, as was clearly demonstrated in Puerto Rico after Hurricane Maria, when low-income, rural, and remote communities suffered significantly longer electricity system outages (Jessel et al., 2019). And these communities are also less frequently able to take advantage of energy programs designed to incentivize energy infrastructure and efficiency upgrades, again for a variety of reasons, including that these programs may require up-front capital costs that low-income households and communities are not able to pay or because their houses are not up to code and thus ineligible. Many such communities also face growing risks from climate change, caused by energy system carbon emissions, which they cannot effectively respond to using only their own resources. These burdens often disproportionately fall on and compound other difficulties faced by communities of color, indigenous communities, low-income and rural communities, people with disabilities, immigrant communities, and other disadvantaged or marginalized groups (Shonkoff et al., 2011; Colon, 2016). From an ethical perspective, this uneven distribution of costs, risks, and benefits—and the unequal power of these communities to self-determination in energy decision-making and to influence energy choices to create fairer and more equitable outcomes—is unjustified and presents a significant opportunity to leverage a clean energy transition to create more just futures for these communities.

Public support for rapid decarbonization: Decarbonization is likely to be among the largest and most significant social, economic, and infrastructural transformations in human history. Public support for such transformation will require securing broad and inclusive agreements across diverse communities with deep stakes in both the present and future of energy systems. Clear knowledge, recognition, and acknowledgement will be essential regarding who has been poorly served by energy systems in the past and present, who will be impacted negatively by energy systems in the future, who will pay for clean energy technologies, who will benefit from them, who has ownership and control over energy systems, and whose voices are given space, recognition, and influence in energy planning and decision-making. Failure has the potential to leave diverse communities either unengaged or in active opposition, undermining commitment to the scale and pace of change required to address rapidly escalating climate risks.

Aspirational foundations of a good society: Energy has been essential to multiple, historical transformations that have significantly improved the human condition. If well designed, the adoption of clean energy technologies and the associated energy systems reconfigurations that it will bring about present a similar opportunity to advance social and technological change in ways that continue to improve wellbeing and thriving.

In a number of ways, renewable energy technologies are well positioned to make future energy systems less damaging—for example, by helping to undo the extensive environmental and health consequences wrought by the burning of carbon-based fuels for energy, on local to global scales. The broad distribution of solar and wind resources, combined with the low and still falling cost of technologies to capture them, means that many countries and communities will have the potential to own and generate their own energy in the future rather than be dependent on others for critical economic infrastructures and supplies, and concentration of industry, and the associated power and wealth that come with it, will be more difficult.

Low-carbon energy technologies are not a panacea, however, and the potential benefits of a low-carbon energy transition will not come automatically. Rather, they will result only from a purposeful effort to design tomorrow's energy systems—and the societies built on them—in ways that contribute to diverse human goals for sustainability, resilience, and thriving.

To accomplish this goal will require significant improvements in research into the social drivers, dynamics, and outcomes of energy transitions and into the relationships between energy systems and human systems, as well as the improved integration of this knowledge into energy planning and system design and implementation. U.S. federal agencies and national laboratories should therefore invest substantially in growing the national capacity to understand the human and social dimensions of energy systems and to assess, visualize, and model their dynamics and structures.

This research should pay special attention to considerations of equity and inequality in existing and future energy systems design and operations. Significant new investments will be needed to analyze and assess the complex dynamic relationships between energy and economic insecurity and the differential implications of energy transitions and systems for a wide variety of communities disadvantaged by existing energy systems and policies; to measure the social and economic outcomes of transition plans and their distribution across different groups; to develop strategies and frameworks for improving the inclusiveness of energy decision making, including especially through improving the effectiveness of community engagement and participation methods; to develop effective strategies of knowledge and policy co-production with diverse communities for the energy system to enhance the relevance and impact of research for communities and decision makers; and to ensure the effectiveness and accountability of strategies for leveraging energy innovation to enhance community economic and social wellbeing and sustainability.

Support Communities, Businesses, and Workers Directly Affected by Transition

Policies and practices in the transition to a low-carbon economy and energy systems should provide significant support for communities, businesses, and workers throughout the United States who will be harmed by and face difficulties as a result of the transition to a carbon-neutral economy. As discussed in the opening of this chapter, the scale and depth of economic transformation anticipated in the economy is large, with the potential to impact a wide swath of communities, businesses, and workers across the nation, in the energy and transportation sectors, and more broadly, who will need considerable help navigating the transition successfully.

Given the implied scale of investment—financial and otherwise—this transformation has the potential to foster sustainable development at multiple scales. Yet, without careful attention to the distribution of costs and benefits, the energy and economic transformation will create, perpetuate, and perhaps even exacerbate highly uneven impacts, with diverse communities bearing concentrated risks and harms, including rural, low-income, communities of color, and other disadvantaged communities (Morello-Frosch et al., 2009). Although systematic research on vulnerabilities to transition impacts is in its infancy in the United States (Carley et al., 2018b; Cha, 2020, 2017; Power et al., 2015), it is widely acknowledged that the impacts of energy system changes will vary geographically and may also be stratified along racial or socioeconomic axes.

In addition, the energy transition necessarily means shifting investment among sectors and industrial activities with direct implications for workers and businesses. Deep decarbonization will result in direct changes to the oil, gas, and coal industries; electric utilities; air, truck, and rail transport; and automobile manufacturing, sales, service, and fueling. It will also require significant changes to the industries that supply parts and equipment, financing, and other support for energy and transportation sectors. Throughout the economy, the disruption of investment and markets for fossil-fuel technologies intersects growing trends in automation and data systems in ways that may amplify losses and challenges for particular groups of workers and businesses, an intersection sometimes referred to as Industry 4.0 (IndustriALL Global Union, 2019).

In areas that host energy infrastructure, both in a concentrated (e.g., coal mines or oil refineries) and more distributed form (e.g., gas stations, grids, or pipelines), as well as corresponding manufacturing facilities, the abandonment or adoption of particular energy technologies and policies directly affect how people support themselves, access healthy environments, and receive essential public services. Experts observe a spiraling fiscal crisis emerging in coal-dependent communities (Morris et al., 2020), which could be replicated in other hot spots absent policy reform. Therefore, deliberate efforts will be required to address the social and economic ruptures created by the energy transition and to secure positive development outcomes in communities and regions.

In many resource-dependent regions, there is a noted temporal and spatial mismatch between jobs lost and jobs created (Power et al., 2015). Here reference is made both to "hot spots" of lost jobs and economic opportunity along supply chains or in sectors made obsolete by the transition to decarbonized energy, such as disruptions to the coal and oil and gas industries and in automobile manufacturing, servicing, and repair owing to the replacement of internal combustion engines with electric motors, as well as to those neighborhoods, cities and towns, and regions that have hosted or will host the industrial-scale facilities associated with manufacturing, generation and storage, and transmission and distribution of low-carbon energy resources. Failure to address these challenges by supporting communities that are confronting them, in an anticipatory and forward-looking manner, has the potential to create new landscapes of economic decline not unlike those of past U.S. industrial transformations and to degrade public support for decarbonization policies.

Defining Support

Ensuring a strong social contract for the transition to a carbon-neutral economy will require identifying the private income and public revenue streams that are lost owing to energy system transformation and generating strategies to replace them. Historically, the nation has benefited financially from its generous fossil fuel mineral endowment, much of which resides on public lands. Many states' public revenues have similarly benefited from the development of energy-resource endowments. In the context of a low-carbon economy, by contrast states' budgetary dependence on revenues from fossil-fuel development revenue for public services and infrastructure acts as a direct barrier to generating the social and political capital necessary to enact systemic change, at least not at the pace demanded for net zero by 2050 (Haggerty, 2018; Mayer, 2018; Haggerty and Haggerty, 2015).

Public policy interventions can help support groups impacted by energy transitions through a variety of mechanisms. Strategies for making policies successful and overcome barriers to success can include, for example, providing direct planning, financial and technical support to affected groups, incentivizing private sector investments, setting rules for markets, and building the capacity of local institutions and communities. Fundamentally, transition support strategies comprise a portfolio of activities focused on identifying and reducing barriers to the ability for workers, business owners, and host communities to pursue self-determined pathways toward sustainable economic activity and new occupational and community identities (e.g., Cha et al., 2019).

When energy infrastructure enables sustainable development in places that host it, it provides dependable sources of private income locally through jobs and other direct payments to individuals, such as leases and royalty payments. More broadly, linking energy systems to sustainable development at local and regional scales hinges on securing public revenue through appropriate taxation that is adequate in timing, amount and form to (1) mitigate any negative impacts of development, (2) encourage the maintenance and stewardship of local environments and services, and (3) encourage long-term economic diversification or other buffers of possible downturns in energy development. Energy system investments can also promote sustainable development through positive synergies between energy development and critical local systems of hard and soft infrastructure. Last, connecting energy system investments to sustainable development means identifying opportunities to leverage and connect across the many nodes in a system to build the capacity of discrete groups and settlements through collective action and investment.

Supporting workers and communities during the transition to a low-carbon energy system involves four goals.

Goal 1: Workers and communities should have accurate information about how clean energy transitions could impact them and should have access to viable economic transition strategies. Uncertainty is a persistent barrier to proactive transition responses by workers and communities. Labor and community leaders report that more information from industry is critical to catalyzing active preparation for lost jobs and other local impacts; however, what public information does exist from elected politicians and facility and mine owners often conveys unfounded optimism or is deliberately obscure. Local government leaders and staff in small, resource-dependent economies (and neighborhoods) tend to lack capacity to generate accurate projections of lost public revenue and its associated social and economic impacts, another circumstance that impedes preparing for transitions (Haggerty et al., 2018; Sanzillo, 2017). Those workers, families, and communities that currently depend directly on fossil fuel-centered activities need a clear message about when and how job losses will occur if they are to be expected to respond in a proactive way.

It is critical to address job losses directly. The isolated nature of coal and other fossil fuel facilities means that they often play an outsized role in local employment.

For example, in Pennsylvania's Greene County, direct employment in coal mining constituted 14 percent of sector employment in 2011. This suggests the importance of offering exit ramps such as early retirement and meaningful job opportunities through retraining and reskilling the fossil fuel workforce—and the importance of the role of labor organizations in advocating for these programs (IndustriALL Global Union, 2019). Policy makers also need to be clear-eyed about the limited opportunities to replace one type of job with another in any given place and the challenges jobs-retraining programs have faced in the past. As one analysis puts it: "The differences in skills and training requirements between these jobs lost [in the coal industry] and jobs gained [in a future clean energy industry] imply the potential for considerable friction in employment in affected communities" (Blue Green Alliance, 2015). In such cases, mobility vouchers (Moretti, 2012) may be practical and realistic responses to transition impact for some workers while frank discussions about rightsizing (though controversial) could benefit local governments. In other geographies, relocation by workers is simply not an option and/or local government services cannot be cut further without drastic consequences such as the loss of public safety resources, libraries, and even basic sanitation. In many cases, sustainable economic development for resource-dependent regions depends on thinking beyond directly replacing one kind of energy employment for another to economic diversification strategies broadly, which is not simple given the dominance of metropolitan regions in the current economy (Goetz et al., 2018). One such example includes jobs in the solar industry, which are growing and outpacing coal jobs (Popovich, 2017); however, coal workers and solar panel workers require different skill sets and there is not necessarily an easy and direct transition. Policy programs and financial incentives that encourage renewable energy development, such as solar, are needed to support the developing market, which includes training workers for jobs in a clean energy economy (Cha, 2017).

Goal 2: Risks to "highly vulnerable" locations where the economic transition to carbon neutrality will exacerbate existing economic disadvantages and health disparities should be directly addressed in transition planning. The association of persistent rural poverty with coal mining in Appalachia is a clear and long-standing example of the risk that dependence on natural resource development can pose for the health and well-being of remote, isolated communities (Lobao et al., 2016; Perdue and Pavela, 2012). So, too, are the issues in fence-line communities or segregated urban neighborhoods dominated by industrial facilities such as power plants and refineries—in these geographies it may not be employment losses, but rather the costs of legacy pollution that compound socioeconomic disadvantages (Cusick, 2020; Plumer et al., 2020). Native American populations experience especially troubling rates of poverty and poor health outcomes as baseline conditions, and these challenges are exacerbated in communities where toxic legacies and job losses from the collapse of energy development also occur. Asking Native American populations to relocate for new jobs conflicts with tribal sovereignty and cultural survival strategies (Wilkinson, 2004). The level of dependence on fossil fuel activity in tribal economies is severe: In one example, the Hopi nation, coal revenues provide 80 percent of the revenues to tribal government (Sanzillo, 2017), and the loss of revenues from the closure of the Navajo Generating Station will severely impact both the Hopi and Navajo nations over the long term, despite efforts by plant owners to address this challenge (Storrow, 2020). A basic goal of any just transition platform is to identify and mitigate these at-risk populations through programs dedicated and tailored to their particular concerns and needs.

Goal 3: Companies should be held accountable for ensuring that fossil fuel energy infrastructures are properly decommissioned and that their long-term environmental impacts are remediated to prevent the creation of persistent environmental contamination and associated health impacts for local populations. Fossil fuel infrastructures are ubiquitous across the U.S. landscape, including wells, pipelines, refineries, storage facilities, and more. Widespread "orphaning" of these infrastructures has the potential to leave many communities facing complex and persistent environmental and health risks from contamination, leakage, and disposal of hazardous materials and equipment. The risks of abandonment without remediation arise with the potential for bankruptcies in the oil, gas, and coal industry associated with decarbonization (Macey and Salovaara, 2019; Walsh, 2017; Walsh and Haggerty, 2017). In the nuclear industry, up-front payments are required into a public investment fund to cover risks of disasters and of decommissioning (NRC, 2019), which could potentially serve as a model for making sure that money is available for decommissioning and remediation of stranded fossil fuel assets. There are also new Economic Development Administration (EDA) nuclear funding options for planning that do not state a sunset for closed plants or a required closure date for open plants to access funds. In FY 2020, $15 million was appropriated to EDA to support communities impacted by nuclear plant closures (EDA, 2020). Eligible affected communities have the opportunity to access these resources and funding in addition to opportunities to consider alternative uses for sites once a nuclear plant decommissioning is complete through other federal programs (EDA, 2019). However, the Price-Anderson Act limits total liability, and power plant owners pay an annual premium per reactor site for $450 million in private insurance for offsite liability coverage, meaning that any large-scale accidents would not be covered by the industry (NRC, 2019). Remediation can provide an important source of local employment (Northern Plains Resource Council, 2018) and in some cases, where appropriate and safe, abandoned facilities may be available for other uses (e.g., as has happened in the case of redevelopment of closed military bases).

States (through their environmental permitting agencies for entities that operate these fossil-fuel development, production, delivery and/or power generation facilities and through their public utility commissions that oversee utility activities, such as integrated resource plans) can play a more active role in requiring such remediation efforts. Similarly, the Federal Energy Regulatory Commission (FERC) could also address such issues with regard to gas-pipeline abandonments. In New Mexico, the Energy Transition Act (ETA) protects consumers and reduces electricity costs as the state moves away from coal and transitions to renewables. The ETA leverages securitization to, in part, provide economic development investment to lessen the local impacts of shutting down a large coal-fired power plant. In the case of PNM's San Juan Generating Station, this mechanism provides over $40 million to assist plant employees, mine workers, and others with severance pay and job training, among other support.

Goal 4: Strategies should be developed to ensure that local, tribal, and state governments are able to replace lost revenue from plant, mine, and other industrial facility closures. As mentioned, local government funding often depends heavily on fossil fuel facilities or extractive activities in areas that host mines and power plants. Outdated fiscal policy plagues resource-dependent regions—tax and expenditure limits adopted during the tax revolt (at both local and state levels) mean that counties cannot grow themselves out of fiscal crisis and that, after decades of extracting valuable natural resources or generating valuable public electricity, they have little to no public funds in reserve to assist with transition. Addressing revenue shortfalls is essential to avoiding further erosion of these communities. One way to redirect revenue would be to require holding back a portion of total federal mineral revenue (which includes bonus payments, rentals, royalties, fines and penalties) and investing it in a permanent endowment from which transition investments can be made.

Reforming and redirecting how fossil fuel revenue is generated and allocated at the national scale will help the United States accomplish three important priorities: (1) weaning the nation off its dependence on fossil fuel for public revenue; (2) establishing a new source of public finance for low-carbon energy infrastructure; and (3) generating funding that is adequate in amount and form to create a realistic source of support for places and businesses affected by transition (Haggerty et al., 2018). Establishing adequate and accessible funds for transition support offers a key signal that the nation honors and respects the contributions of fossil fuels to two centuries of national prosperity.

Rationale

Over the past decade, reductions in coal use in the United States for both industrial use and electricity generation have given rise to a spate of bankruptcies in the industry, closure of mines, and significant decreases in the jobs and resources provided by the industry to the communities in which it operates. The coal industry is modest in size, compared to the U.S. economy, but its concentration in local geographies—for example, in Kentucky and in the Powder River Basin in Wyoming—have contributed to the outsized impact of the industry's decline in those places. The oil and gas industry also has clear areas where it is concentrated, geographically, although these are more widely scattered across diverse regions of the United States, and oil and gas infrastructure is more widely distributed in different parts of the country, meaning that the impacts of declining oil and gas production through the transition to a carbon neutral economy will be more widely felt across the country over the next three decades, especially when combined with declines in the manufacturing of internal combustion engine parts and automobiles. These industries are a much larger fraction of the U.S. economy than coal and provide significant employment, tax revenues, capital expenditures, and infrastructure investments that benefit many regions of the U.S. economy and a large fraction of American communities.

While future energy systems should also provide extensive future benefits to U.S. workers and communities, as well as the U.S. economy, as described above, this transformation is likely to create uneven distributions of costs to different communities. Unless addressed through effective support programs, these costs will include both direct job losses and losses to public revenues, indirect job losses and declines in general business revenues in impacted communities that lose major industries, threats to community and worker identities and happiness, persistent geographies of economic decline, and resentment, anger, and perhaps even opposition to decarbonization.

The kinds of challenges confronting workers, families, communities, and businesses in communities impacted severely by the transition result from both market and policy failures. The Business Roundtable has argued that workers and communities deserve appropriate consideration as stakeholders in business decision making, and thus might expect assistance in economic transitions from declining industries (Gelles and Yaffe-Bellany, 2019).

Most businesses still operate, however, according to decision-making logics that reward and consider only the interests and voices of shareholders, paying little regard to the needs of workers and communities and even at times operating in ways that degrade worker and community capabilities to plan for and execute transition planning. This approach is compounded by market and policy failures that currently undermine the contribution of energy infrastructure to local or regional development opportunities: (1) a persistent scalar mismatch in decision making, where local communities have little input about energy futures, either in corporate decision making or regulatory decision making; (2) fiscal policies that trade incentives to industry against the long-term ability for communities to benefit, coupled to business practices that routinely seek to secure special tax deals; and (3) a failure of both business and policy to anticipate and plan for the end-of-life stage of major industrial systems.

Ensuring that policy and its implementation deliberately link system designs and regional and local priorities makes sense for two fundamental reasons. The first is that new energy infrastructure is more likely to win public support when proposed projects have demonstrable social and economic benefits in host areas (Boudet, 2019)—thereby reducing, although hardly eliminating, costly delays and resistance to unpopular projects. It is important that, as early as possible, the stages of new energy infrastructure include decommissioning plans so that the public has faith that their communities will not be left paying for or living with infrastructure when it becomes obsolete. The second is that without policy reform, the energy infrastructure transitions associated with the transition to a carbon neutral economy will exacerbate development challenges rather than benefits in many places that host energy projects. Local planning would also be significantly enhanced by the inclusion of capacity building for workforce and community transitions within corporate social responsibility and sustainability metrics for the sectors of the economy facing major economic transformations, including oil and gas, automobile manufacturing, and the financial sector. Encouraging affected communities and workforces to acknowledge and take seriously the challenges posed by decarbonization could help significantly improve success rates, especially if launched well before facility closures.

The United States has built the world's largest and most successful energy and automobile industries and infrastructures. The transformation of these economic sectors will leave a historic legacy of challenges for communities that have benefited historically from the exploitation of carbon-based fuels. The geographically concentrated wealth generated by carbon mining are difficult to replicate in the more distributed solar, wind, and battery industries. This has the potential to significantly reduce inequalities driven by that concentration of wealth, but it will also create disruptive effects in communities facing that decline that for most places will not be able to be replaced on a one-to-one basis. It may also raise questions about how to equitably share the costs and burdens of diverse assets that are stranded as carbon-based infrastructures are closed. Diverse policy rules that have further benefited those communities will, in turn, compound harms, for example, state laws that pay communities higher rates for transmission lines that carry coal-fired electricity versus newer renewables. There are already other challenges confronting these communities, such as abandoned coal mines and, in the future, abandoned oil and gas infrastructure unless appropriate steps are taken now to anticipate and proactively address these problems (Partridge et al., 2020).

Maximize Cost-Effectiveness

The final criterion for evaluating pathways to a future carbon-neutral U.S. economy is that policies to support the transition and ensure that it enhances U.S. economic strength, promotes equity and inclusion, and supports communities should be accomplished in as cost-effective a manner as possible. The economic investment required to transition the U.S. economy to carbon neutrality will be extensive and will require widespread coordination among the diverse sectors and actors necessary. The scale of the required investment is large enough that it will impinge on other national priorities and on the overall economy.

Committing to cost-effectiveness as a core criterion for evaluating pathways, alongside the other goals identified in Chapter 3, ensures that policies to advance carbon neutrality are achieved at the lowest possible overall costs and prioritizes investments and policies that create flexibility in how goals are achieved that allow for cost reductions wherever possible. Prioritizing effective investments, therefore, works to bolster the social contract for the U.S. transition to carbon neutrality by maximizing the impact of each investment and by lowering political opposition tied to concerns about costs and regulation. Cost-effectiveness should not be applied, however, as the sole criterion for consideration, nor be focused solely on carbon emissions reductions. The goal is not to find the most cost-effective strategies to reduce carbon emissions but rather to find the most cost-effective strategies to reduce carbon emissions while also strengthening the U.S. economy, promoting inclusion and equity, and supporting communities facing transitions.

Defining Cost-Effectiveness

Cost-effectiveness, as traditionally used by environmental economists, refers to the idea of achieving a given environmental or social outcome at the lowest aggregate cost to society (e.g., Hahn and Stavins, 1992). Here, aggregate cost refers to the societal resources diverted to comply with a particular policy, and equivalently the goods and services foregone by households and/or the government as that diversion occurs, versus a counterfactual absent the policy, and without regard to who bears those costs. It does not include any environmental impacts.

Cost-effectiveness is often contrasted with what economists refer to as efficiency, which does not take the environmental outcome as given but instead seeks to maximize aggregate net benefits (National Center for Environmental Economics, 2014). In other words, efficiency calculations assess what policy maximizes the monetized environmental benefits minus the aggregate cost noted above. This can include a variety of different benefits, such as reductions in health costs from pollution or inclusion of the social cost of carbon. Again, this is without regard to who bears the costs or receives the benefits.

Rationale for Cost-Effectiveness and Other Considerations

There is a long history and debate over the role of cost-benefit analysis in policy analysis, but most would argue it remains a useful metric (Arrow et al., 1996, p. 221). Seeking to understand the aggregate costs and benefits, among other criteria, is necessary to appreciate where society should spend scarce resources.

Measuring environmental, mortality, and morbidity benefits creates particular ethical dilemmas and analytic difficulties (Jamison et al., 2006). As opposed to efficiency, or maximizing net benefits, cost-effectiveness has the advantage of not requiring such an effort. In the present context, where the case for carbon neutrality is already made, focusing on the aggregate benefits of decarbonization itself at this stage is unnecessary.

Focusing on the aggregate economic costs of proposed policies, and seeking to lower them, ensures that resources remain available to tackle other social problems as well as promoting economic well-being. At the same time, it has long been recognized that lowering aggregate costs to society often comes at the expense of achieving equitable costs across members of society (Okun, 1974). Thus, cost-effectiveness must be considered only alongside other criteria, using multiple-criteria methods and negotiation frameworks among politicians, stakeholders, and the public that allow for consideration not only of the cost-effectiveness of achieving a net-zero carbon economy but also the aggregate co-benefits or externalities of different policies, as well as the distribution of costs and benefits across groups. The committee recognizes that these other goals—regarding the economy, equity, and transition—are not as precisely defined and place constraints on cost-effectiveness.

In addition, the pathways to decarbonization discussed in Chapter 2 anticipate a combination of multiple changes to the economy and energy system, and the policies recommended in Chapter 4 are also meant to be adopted as a package. This can make traditional methods of evaluating cost-effectiveness less accurate, if calculations are done for each individual policy, independently, because policies adopted as a package may have interactive effects that either reduce or enhance cost-effectiveness in comparison to the same policies adopted separately. Cost-effectiveness and other criteria thus need to be evaluated for the program as a whole, not just individual parts.

Cost-effectiveness as a criterion for policy design frequently points to flexibility in compliance (Schmalensee and Stavins, 2017). For example, rather than requiring a particular technology to achieve an environmental outcome, such as zero-carbon emissions, define the performance requirement and leave firms and households free to achieve the goal however they wish. A staple of these programs has been emissions or credit trading programs of one sort or another. The acid rain trading program is widely acknowledged to have significantly reduced sulfur dioxide emissions at a significantly lower cost than likely alternatives (Carlson et al., 2000). Individual coal-fired power plants were given limited allocations of emission allowances, which they needed to surrender annually, one-for-one, for each ton of sulfur dioxide that they emitted. Firms with higher emissions could purchase additional allowances from other firms who overcomplied. However, it may be difficult to consider cost-effectiveness for a large-scale transformation like the transition to electric vehicles, as noted in the 2013 report, *Transitions to Alternative Vehicles and Fuels* (National Research Council, 2013). The report attempts to look at policies, costs, and benefits to reduce greenhouse gases by 80 percent by 2050, while attempting to include transition barriers such as resistance to novel products, lack of infrastructure, lack of choice diversity, economies of scale, learning by doing, and the time constraints for change as well as interactions with the fuel-electricity systems. However, the committee that authored this earlier report found it difficult to estimate the most cost-effective pathway to do such a transition.

A frequent concern with emissions trading and other market-based policies is that they can create hot spots where emissions persist and that the environmental (and economic) consequences may be inequitable. For example, Ringquist (2011) argues that the acid rain trading program did not concentrate emissions in Black or Hispanic communities, but did concentrate emissions in poorly educated communities. Similarly, environmental justice advocates anticipated, warned about, and ultimately documented hot spots from air toxics and criteria pollutants resulting from California's greenhouse gas emissions trading program that required follow up policies to reduce inequitable impacts (Cushing et al., 2018). In contrast, a recent study by Hernandez-Cortes and Meng (2020) suggests that the California program reduced the pollution exposure gap among communities.

Perhaps more relevant to many of our recommendations focused on performance standards, the lead phasedown program in the United States effectively removed lead from gasoline at a significant cost savings (Newell and Rogers, 2006). In the lead phasedown, refineries were allowed a certain declining concentration of lead in refined gasoline during the 1980s. To the extent they over- or underachieved the target, they calculated the total volume of excess- or under-emitted lead. Those excess-emitting refiners were required to buy that amount of credits from under-emitting refiners; credits could also be banked for use in future years. By ratcheting down the benchmark for compliance, lead was effectively eliminated from gasoline.

As an analogy to the roadmap for creating a carbon neutral economy, seeking to eliminate fossil fuel emissions in the same way lead was eliminated from gasoline creates a risk of equity concerns in potential hot spots. Another concern with policies that provide for flexibility designed to foster cost-effectiveness arises if the lowest-cost strategies either result in lower co-benefits or higher externalities or ancillary risks. For example, carbon capture and sequestration and renewable technologies have very different co-benefits and risks with regard to environmental pollution and health effects, impacts on the electricity grid, and so on.

Given the disruption of traditional energy systems, markets, and workforces anticipated with decarbonization policies, these secondary benefits and costs may be substantially different and should be seriously considered in evaluating policies. Thus, cost-effectiveness exists as only one of several ends toward which this deep decarbonization plus framework drives. The establishment and maintenance of a social contract for a national low-carbon economic and energy transition demand attention and consideration for the full array of implications of policy choices for the economy and society.

REFERENCES

Allison, G. 2019. "Is China Beating America to AI Supremacy?" The National Interest. https://nationalinterest.org/feature/china-beating-america-ai-supremacy-106861.

Arrow, K.J., M.L. Cropper, G.C. Eads, R.W. Hahn, L.B. Lave, R.G. Noll, P.R. Portney, et al. 1996. Is there a role for benefit-cost analysis in environmental, health, and safety regulation? *Science* 272(5259): 221–222.

Ashworth, P., A. Littleboy, P. Graham, and S. Niemeyer. 2011. "Turning the Heat On: Public Engagement in Australia's Energy Future." Pp. 131-148 in *Renewable Energy and the Public* (P. Devine-Wright, ed.). New York, NY: Taylor & Francis.

Baptista, A.I., and A. Perovich. 2020. *Environmental Justice and Philanthropy: Challenges and Opportunities for Alignment Gulf South and Midwest Case Studies*. Building Equity and Alignment for Environmental Justice. https://bea4impact.org.

Bednar, D.J., and T.G. Reames. 2020. Recognition of and response to energy poverty in the United States. *Nature Energy* 5(6): 1–8.

Berkes, F. 2009. Evolution of co-management: Role of knowledge generation, bridging organizations and social learning. *Journal of Environmental Management* 90(5): 1692–1702.

Bice, S., and T.B. Fischer. 2020. Impact assessment for the 21st century—What future? *Impact Assessment and Project Appraisal* 38(2): 89–93.

Blackburn, G., C. Magee, and V. Rai. 2014. Solar valuation and the modern utility's expansion into distributed generation. *The Electricity Journal* 27(1): 18–32.

Blue Green Alliance. 2015. *America's Energy Transition: A Case Study in the Past and Present of Southwestern Pennsylvania's Power Sector.* https://www.bluegreenalliance.org.

Boroush, M. 2020. "Research and Development: U.S. Trends and International Comparisons." National Science Board. https://ncses.nsf.gov/pubs/nsb20203/executive-summary.

Boudet, H.S. 2019. Public perceptions of and responses to new energy technologies. *Nature Energy* 4(6): 446–455.

Bridge, G., B. Özkaynak, and E. Turhan. 2018. Energy infrastructure and the fate of the nation: Introduction to special issue. *Energy Research & Social Science* 41: 1–11.

Bullard, R.D. 2015. Environmental justice in the 21st century: Race still matters. *Phylon (1960–)* 52(1): 72–94.

Burdge, R.J. 2002. Why is social impact assessment the orphan of the assessment process? *Impact Assessment and Project Appraisal* 20(1): 3–9.

Burger, S.P., and M. Luke. 2017. Business models for distributed energy resources: A review and empirical analysis. *Energy Policy* 109: 230–248.

Bush, M.J. 2019. Denial and Deception. Pp. 373–420 in *Climate Change and Renewable Energy*. Palgrave Macmillan, Cham.

Byrne, J., and Y.M. Mun. 2003. *Rethinking Reform in the Electricity Sector: Power Liberalisation or Energy Transformation?* https://jbyrne.org.

California Energy Commission. 2018. *Energy Equity Indicators Tracking Progress.* Sacramento, CA.

Carley, S., and D.M. Konisky. 2020. The justice and equity implications of the clean energy transition. *Nature Energy* 5(8): 569–577.

Carley, S., T.P. Evans., and D.M. Konisky. 2018a. Adaptation, culture, and the energy transition in American coal country. *Energy Research & Social Science* 37: 133–139.

Carley, S., T.P. Evans, M. Graff, and D.M. Konisky. 2018b. A framework for evaluating geographic disparities in energy transition vulnerability. *Nature Energy* 3(8): 621–627.

Carlson, C., D. Burtraw, M. Cropper, M., and K.L. Palmer. 2000. Sulfur dioxide control by electric utilities: What are the gains from trade? *Journal of Political Economy* 108(6): 1292–1326.

Cha, J.M. 2017. A just transition: Why transitioning workers into a new clean energy economy should be at the center of climate change policies. *Fordham Environmental Law Review* 29(2): 196–220.

Cha, J. M. 2020. A just transition for whom? Politics, contestation, and social identity in the disruption of coal in the Powder River Basin. *Energy Research and Social Science* 69.

Cha, J. M., M. Pastor, M. Wander, J. Sadd, and R. Morello-Frosch. 2019. *A Roadmap to an Equitable Low-Carbon Future: Four Pillars for a Just Transition.* The Climate Equity Network. Los Angeles, CA: Program for Environmental and Regional Equity, University of Southern California. https://dornsife.usc.edu/pere/roadmap-equitable-low-carbon-future.

Chu, S., and A. Majumdar. 2012. Opportunities and challenges for a sustainable energy future. *Nature* 488(7411): 294–303.

City of Minneapolis. 2013. *Minneapolis Climate Action Plan.* Minneapolis, MN: Minneapolis City Coordinator.

City of Portland. 2020. "About PCEF." https://www.portland.gov/bps/cleanenergy/about-pcef.

City of Providence. 2019. *The City of Providence's Climate Justice Plan.* Providence, RI.

Colon, J. 2016. *The Disproportionate Impacts of Climate Change on Communities of Color in Washington State.* Seattle, WA: Front and Centered.

Coumarianos, J., and L.P. Norton. 2020. "BlackRock Passes a Milestone, With $7 Trillion in Assets Under Management." *Barron's.* https://www.barrons.com/articles/blackrock-earnings-assets-under-management-7-trillion-51579116426.

CPUC (California Public Utility Commission). 2019. "Proceedings/Programs Targeting Disadvantaged Communities (DACs) or Low-Income Households." https://www.cpuc.ca.gov/uploadedFiles/CPUCWebsite/Content/UtilitiesIndustries/Energy/EnergyPrograms/Infrastructure/DC/20190614_DAC%20Programs%20w%20proceedings(1).pdf.

CRS (Congressional Research Service). 2020. "U.S. Research and Development Funding and Performance: Fact Sheet." https://fas.org/sgp/crs/misc/R44307.pdf.

Cushing, L., D. Blaustein-Rejto, M. Wander, M. Pastor, J. Sadd, A. Zhu, and R. Morello-Frosch. 2018. Carbon trading, co-pollutants, and environmental equity: Evidence from California's Cap-and-Trade Program (2011–2015). *PLOS Medicine* 15(7).

Cusick, D. 2020. "Past Racist 'Redlining' Practices Increased Climate Burden on Minority Neighborhoods." *E&E News*. https://www.scientificamerican.com/article/past-racist-redlining-practices-increased-climate-burden-on-minority-neighborhoods/.

Dao, E. 2020 "Fighting Climate Change Isn't Just an Environmental Issue—It's a Social Justice Issue Too." *The Rising*. https://therising.co/2020/06/04/environmental-justice-social-equity/.

Devine-Wright, P. 2011. Public engagement with large-scale renewable energy technologies: Breaking the cycle of NIMBYism. *Wiley Interdisciplinary Reviews: Climate Change* 2(1): 19–26.

Dietz, T., G.T. Gardner, J. Gilligan, P.C. Stern, and M.P. Vandenbergh. 2009. Household actions can provide a behavioral wedge to rapidly reduce US carbon emissions. *Proceedings of the National Academy of Sciences U.S.A.* 106(44): 18452–18456.

DOE-SETO (U.S. DOE Solar Energy Technologies Office). 2020. *2020 Peer Review Report*. Washington DC: Office of Energy Efficiency and Renewable Energy.

Drehobl, A., L. Ross, and R. Ayala. 2020. *How High Are Household Energy Burdens?* Washington DC: American Council for an Energy-Efficient Economy.

Dunlap, A. 2017. 'A bureaucratic trap:' Free, prior and informed consent (FPIC) and wind energy development in Juchitán, Mexico. *Capitalism Nature Socialism* 29(4): 88–108.

Dwyer, J., and D. Bidwell. 2019. Chains of trust: Energy justice, public engagement, and the first offshore wind farm in the United States. *Energy Research and Social Science* 47: 166–176.

EDA (U.S. Economic Development Administration). 2019. *Report on Best Practices to Assist Communities Affected By Loss of Tax Revenue and Job Loss Due to Nuclear Power Plant Closures*. Washington DC: U.S. Department of Commerce.

EDA. 2020. "EDA Seeks Applications to Support Nuclear Closure Communities." https://www.eda.gov/news/blogs/2020/05/11/Nuclear-Closure-Communities.htm.

Energy Entrepreneurs. 2020. *Clean Jobs America 2020*. https://e2.org/wp-content/uploads/2020/04/E2-Clean-Jobs-America-2020.pdf.

EPA (Environmental Protection Agency). n.d. "Environmental Justice." https://www.epa.gov/environmentaljustice.

Esteves, A.M., D. Franks, and F. Vanclay. 2012. Social impact assessment: The state of the art. *Impact Assessment and Project Appraisal* 30(1): 34–42.

European Commission. 2020. "Just Transition Platform." https://ec.europa.eu/info/strategy/priorities-2019-2024/european-green-deal/actions-being-taken-eu/just-transition-mechanism/just-transition-platform_en.

Ewing, J. 2019. "Volkswagen Moves to Rapidly Increase Production of Electric Cars." *The New York Times*. March 12. https://www.nytimes.com/2019/03/12/business/volkswagen-electric-cars.html.

Farrell, J. 2019. The growth of climate change misinformation in US philanthropy: Evidence from natural language processing. *Environmental Research Letters* 14(3).

Farrell, J., K. McConnell, R. and Brulle. 2019. Evidence-based strategies to combat scientific misinformation. *Nature Climate Change* 9: 191–195.

Ferris, J.M., and N.P.O. Williams. 2012. *Philanthropy and Government Working Together: The Role of Offices of Strategic Partnerships in Public Problem Solving*. Los Angeles, CA: Center on Philanthropy and Public Policy, University of Southern California.

Finley-Brook, M., and E. Holloman. 2016. Empowering energy justice. *International Journal of Environmental Research and Public Health* 13(9): 926.

Firestone, J., B. Hoen, J. Rand, D. Elliott, G. Hübner, and J. Pohl. 2017. Reconsidering barriers to wind power projects: Community engagement, developer transparency and place. *Journal of Environmental Policy and Planning* 20(3): 370-386.

Firestone, J., C. Hirt, D. Bidwell, M. Gardner, and J. Dwyer. 2020. Faring well in offshore wind power siting? Trust, engagement and process fairness in the United States. *Energy Research and Social Science* 62.

Fleming, K.L. 2018. Social equity considerations in the new age of transportation: Electric, automated, and shared mobility. *Journal of Science Policy and Governance* 13(1).

Fowlie, M., R. Walker, and D. Wooley. 2020. *Climate Policy, Environmental Justice, and Local Air Pollution*. Washington DC: Brookings Institution.

Funkhouser, E., G. Blackburn, C. Magee, and V. Rai. 2015. Business model innovations for deploying distributed generation: The emerging landscape of community solar in the U.S. *Energy Research and Social Science* 10: 90–101.

Gelles, D., and D. Yaffe-Bellany. 2019. "Shareholder Value is No Longer Everything, Top C.E.O.s Say." *The New York Times*. August 19. https://www.nytimes.com/2019/08/19/business/business-roundtable-ceos-corporations.html.

Giordono, L.S., H.S. Boudet, A. Karmazina, C.L. Taylor, and B.S. Steel. 2018. Opposition "overblown"? Community response to wind energy siting in the Western United States. *Energy Research and Social Science* 43: 119–131.

Goetz, S.J., M.D. Partridge, and H.M. Stephens. 2018. The economic status of rural America in the President Trump era and beyond. *Applied Economic Perspectives and Policy* 40(1): 97–118.

Gordon, R.J. 2016. *The Rise and Fall of American Growth: The U.S. Standard of Living since the Civil War*. The Princeton Economic History of the Western World. Princeton, NJ: Princeton University Press.

Haggerty, J.H., M.N. Haggerty, K. Roemer, and J. Rose. 2018. Planning for the local impacts of coal facility closure: Emerging strategies in the U.S. West. *Resources Policy* 57: 69-80.

Haggerty, M.N. 2018. Rethinking the fiscal relationship between public lands and public land counties: County payments 4.0. *Humboldt Journal of Social Relations* 1(40): 116–136.

Haggerty, M.N., and J.H. Haggerty. 2015. *An Assessment of U.S. Federal Coal Royalties: Current Royalty Structure, Effective Royalty Rates, and Reform Options*. Bozeman, MT: Headwaters Economics.

Hahn, R.W., and R.N. Stavins. 1992. Economic incentives for environmental protection: Integrating theory and practice. *The American Economic Review* 82(2): 464–468.

Haines, A. 2017. Health co-benefits of climate action. *The Lancet Planetary Health* 1(1): 4–5.

Hall, N., P. Ashworth, and P. Devine-Wright. 2013. Societal acceptance of wind farms: Analysis of four common themes across Australian case studies. *Energy Policy* 58: 200–208.

Henry, M.S., M.D. Bazilian, and C. Markuson. 2020. Just transitions: Histories and futures in a post-COVID world. *Energy Research and Social Science* 68.

Hernandez-Cortes, D., and K.C. Meng. 2020. "Do Environmental Markets Cause Environmental Injustice? Evidence from California's Carbon Market." Working Paper 27205. National Bureau of Economic Research. https://www.nber.org.

Hirsch, T., M. Matthess, and J. Funfgelt. 2017. *Guiding Principles and Lessons Learnt for a Just Energy Transition in the Global South*. Bonn, Germany: Friedrich-Ebert-Stiftung.

Howe, P.D., M. Mildenberger, J.R. Marlon, and A. Leiserowitz. 2015. Geographic variation in opinions on climate change at state and local scales in the USA. *Nature Climate Change* 5(6): 596–603.

IEA (International Energy Agency). 2020a. "Clean Energy Technology Guide." https://www.iea.org/articles/etp-clean-energy-technology-guide.

IEA. 2020b. *Special Report on Clean Energy Innovation*. Paris, France.

IndustriALL Global Union. 2019. "A Just Transition for Workers." http://www.industriall-union.org/a-just-transition-for-workers.

International Labour Organization. 2016. "Guidelines for a Just Transition Towards Environmentally Sustainable Economies and Societies for All." https://www.ilo.org/global/topics/green-jobs/publications/WCMS_432859/lang—en/index.htm.

IRENA (International Renewable Energy Agency). 2019. *Global Energy Transformation: A Roadmap to 2050*. Abu Dhabi.

Jacquet, J.B., and R.C. Stedman. 2013. The risk of social-psychological disruption as an impact of energy development and environmental change. *Journal of Environmental Planning and Management* 57(9): 1285–1304.

Jamison, D.T., J.G. Breman, A.R. Measham, G. Alleyne, M. Claeson, D.B. Evans, P. Jha, et al. 2006. *Disease Control Priorities in Developing Countries*. New York, Washington, DC: Oxford University Press; World Bank.

Jenkins, J.D., D. Swezey, and Y. Borofsky. 2010. "Where Good Technologies Come From: Case Studies In American Innovation." The Breakthrough Institute. https://thebreakthrough.org/articles/american-innovation.

Jenkins, J.M., D.A. Caldwell, H. Chandrasekaran, J.D. Twicken, S.T. Bryson, E.V. Quintana, B.D. Clarke, et al. 2010. Overview of the Kepler Science Processing Pipeline. *The Astrophysical Journal* 713(2): 87–91.

Jessel, S., S. Sawyer, and D. Hernández. 2019. Energy, poverty, and health in climate change: A comprehensive review of an emerging literature. *Frontiers in Public Health* 7: 357.

Just Transition Fund. 2020. "The National Economic Transition Platform." https://nationaleconomictransition.org/.

Kimura, A.H., and A. Kinchy. 2019. *Science by the People: Participation, Power, and the Politics of Environmental Knowledge.* New Brunswick, NJ: Rutgers University Press.

Krukowska, E., and N. Chrysoloras. 2019. "Europe Set to Overhaul Its Entire Economy in Green Deal Push." Bloomberg. https://www.bloomberg.com/news/articles/2019-12-02/europe-set-for-green-deal-that-will-radically-change-its-economy.

Ladislaw, S., and J. Barnet. 2019. *Energy in America: The Changing Role of Energy in the U.S. Economy.* Washington, DC: Center for Strategic and International Studies.

Leiserowitz, A.A., E.W. Maibach, C. Roser-Renouf, N. Smith, and E. Dawson. 2013. Climategate, public opinion, and the loss of trust. *American Behavioral Scientist* 57(6): 818–837.

Leiserowitz, A., E. Maibach, S. Rosenthal, J. Kotcher, A. Gustafson, P. Bergquist, M. Goldberg, et al. 2018. *Energy in the American Mind, December 2018.* New Haven, CT: Yale Program on Climate Change Communication.

Lerza, C. 2011. *A Perfect Storm—Lessons from the Defeat of Proposition 23.* Funders Network on Transforming the Global Economy. https://edgefunders.org.

Lester, R.K., and D.M. Hart. 2012. *Unlocking Energy Innovation: How America Can Build a Low-Cost, Low-Carbon Energy System.* Cambridge, MA: MIT Press.

Lewandowsky, S., U.K. Ecker, and J. Cook. 2017. Beyond misinformation: Understanding and coping with the "post-truth" era. *Journal of Applied Research in Memory and Cognition* 6(4): 353–369.

Lewis, J.I. 2014. Industrial policy, politics and competition: Assessing the post-crisis wind power industry. *Business and Politics* 16(4): 511–547.

Liévanos, R. 2018. Retooling CalEnviroScreen: Cumulative pollution burden and race-based environmental health vulnerabilities in California. *International Journal of Environmental Research and Public Health* 15(4): 762.

Lobao, L., M. Zhou, M. Partridge, and M. Betz. 2016. Poverty, place, and coal employment across Appalachia and the United States in a new economic era. *Rural Sociology* 81(3): 343-386.

Macey, J., and J. Salovaara. 2019. Bankruptcy as bailout: Coal company insolvency and the erosion of federal law. *Stanford Law Review* 71: 879-958.

Madlener, R. 2020. "Sustainable Energy Transition and Increasing Complexity: Trade-Offs, the Economics Perspective and Policy Implications." Pp. 251–286 in *Inequality and Energy*. Academic Press.

Mayer, J. 2018. "How U.S. Cities Can Finance Resilient and Equitable Infrastructure." *American City and County.* https://www.americancityandcounty.com/2018/04/11/how-u-s-cities-can-finance-resilient-and-equitable-infrastructure/.

McCollum, D.L., C. Wilson, M. Bevione, S. Carrara, O.Y. Edelenbosch, J. Emmerling, C. Guivarch, et al. 2018. Interaction of consumer preferences and climate policies in the global transition to low-carbon vehicles. *Nature Energy* 3(8): 664–673.

McLaren Loring, J. 2007. Wind energy planning in England, Wales and Denmark: Factors influencing project success. *Energy Policy* 35(4): 2648–2660.

Mercer, N., A. Hudson, D. Martin, and P. Parker. 2020. That's our traditional way as indigenous peoples: Towards a conceptual framework for understanding community support of sustainable energies in NunatuKavut, Labrador. *Sustainability* 12(15).

Miller, C.A., A. Iles, and C.F. Jones. 2013. The social dimensions of energy transitions. *Science as Culture* 22(2): 135–148.

Miller, C.A., J. Richter, and J. O'Leary. 2015. Socio-energy systems design: A policy framework for energy transitions. *Energy Research and Social Science* 6: 29–40.

Morello-Frosch, R., M. Pastor, J. Sadd, and S. Shonkoff. 2009. *The Climate Gap: Inequalities in How Climate Change Hurts Americans and How to Close the Gap.* Los Angeles, CA: Program for Environmental and Regional Equity, University of Southern California.

Moretti, E. 2012. *The New Geography of Jobs.* Boston, MA: Houghton Mifflin Harcourt.

Morris, A., N. Kaufman, and S. Doshi. 2020. "Revenue at Risk in Coal-Reliant Counties." *Environmental and Energy Policy and the Economy.* University of Chicago Press.

Mulvaney, D. 2014. Are green jobs just jobs? Cadmium narratives in the life cycle of photovoltaics. *Geoforum; Journal of Physical, Human, and Regional Geosciences* 54: 178–186.

Narassimhan, E., K.S. Gallagher, S. Koester, and J.R. Alejo. 2018. Carbon pricing in practice: A review of existing emissions trading systems. *Climate Policy* 18(8): 967–991.

NASEM (National Academies of Sciences, Engineering, and Medicine). 2019. *Deployment of Deep Decarbonization Technologies: Proceedings of a Workshop*. Washington, DC: The National Academies Press.

NASEO (National Association of State Energy Officials) and EFI (Energy Futures Initiative). 2020. *The 2020 U.S. Energy and Employment Report 2020*. https://www.usenergyjobs.org.

National Center for Environmental Economics. 2014. *Guidelines for Preparing Economic Analyses*. Washington DC: Environmental Protection Agency.

National Research Council. 2008. *Public Participation in Environmental Assessment and Decision Making*. Washington, DC: The National Academies Press.

National Research Council. 2013. *Transitions to Alternative Vehicles and Fuels*. Washington, DC: The National Academies Press.

Nemet, G.F., M. Jakob, J.C. Steckel, and O. Edenhofer. 2017. Addressing policy credibility problems for low-carbon investment. *Global Environmental Change: Human and Policy Dimensions* 42: 47–57.

Nemet, G.F. 2019. *How Solar Energy Became Cheap: A Model for Low-Carbon Innovation*. London: New York: Routledge, Taylor and Francis Group.

Newell, P., and D. Mulvaney. 2013. The political economy of the "just transition." *The Geographical Journal* 179(2): 132–140.

Newell, R., and K. Rogers. 2006. "The Market-Based Lead Phasedown." *Moving to Markets in Environmental Regulation*. Oxford University Press.

Northern Plains Resource Council. 2018. *Doing it Right: Colstrip's Bright Future with Cleanup*. Billings, MT.

NRC (Nuclear Regulatory Commission). 2019. "Backgrounder on Nuclear Insurance and Disaster Relief." https://www.nrc.gov/reading-rm/doc-collections/fact-sheets/nuclear-insurance.html.

O'Brien, K., B. Hayward, and F. Berkes. 2009. Rethinking social contracts: Building resilience in a changing climate. *Ecology and Society* 14(2).

Okun, A.M. 1974. "Unemployment and Output in 1974." Brookings Papers on Economic Activity, No. 2, 495-506.

Oppel Jr., R.A., R. Gebeloff, K.K.R. Lai, W. Wright, and M. Smith. 2020. "The Fullest Look Yet at the Racial Inequity of Coronavirus." *The New York Times*. July 4.

Papillon, M., and T. Rodon. 2017. Proponent-indigenous agreements and the implementation of the right to free, prior, and informed consent in Canada. *Environmental Impact Assessment Review* 62: 216–224.

Partridge, T., J. Barandiaran, C. Walsh, K. Bakardzhieva, L. Bronstein, and M. Hernandez. 2020. California Oil: Bridging the gaps between local decision-making and state-level climate action. *The Extractive Industries and Society* 7(4): 1354-1359.

Perdue, R.T., and G. Pavela. 2012. Addictive economies and coal dependency: Methods of extraction and socioeconomic outcomes in West Virginia, 1997-2009. *Organization and Environment* 25(4): 368–384.

Petrova, M.A. 2013. Public acceptance of wind energy in the United States: NIMBYism Revisited. *Wiley Interdisciplinary Reviews: Climate Change* 4(6): 575–601.

Pezzullo, P.C., and R. Cox. 2017. *Environmental Communication and the Public Sphere*. Thousand Oaks, CA: SAGE Publications.

Platzer, M.D. 2012. *US Solar Photovoltaic Manufacturing: Industry Trends, Global Competition, Federal Support*. Washington DC: Congressional Research Service.

Plumer, B., N. Popovich, and M. Renault. 2020. "How Racist Urban Planning Left Some Neighborhoods to Swelter." *The New York Times*. August 26. https://www.nytimes.com/2020/08/26/climate/racist-urban-planning.html.

Popovich, N. 2017. "Today's Energy Jobs are in Solar, Not Coal." *The New York Times*. April 25. https://www.nytimes.com/interactive/2017/04/25/climate/todays-energy-jobs-are-in-solar-not-coal.html.

Porter, S.E., and K. Hardin. 2020. "Navigating the Energy Transition from Disruption to Growth." Deloitte. https://www2.deloitte.com/us/en/insights/industry/power-and-utilities/future-of-energy-us-energy-transition.html.

Power, M.W., M.G. Mosley, J. Frankel, and P. Hanser. 2015. *Managing the Employment Impact of Energy Transition in Pennsylvania Coal Country*. BlueGreen Alliance.

Rai, V., and A.L. Beck. 2015. Public perceptions and information gaps in solar energy in Texas. *Environmental Research Letters* 10(7).

Ramos, T., and S.M. Pires. 2013. "Sustainability Assessment: The Role of Indicators." Pp. 81–99 in *Sustainability Assessment Tools in Higher Education Institutions*. Cham: Springer International Publishing.

Renn, O., F. Ulmer, and A. Deckert. 2020. *The Role of Public Participation in Energy Transitions*. Cambridge, MA: Academic Press.

Ringquist, E.J. 2011. Trading equity for efficiency in environmental protection? Environmental justice effects from the SO$_2$ Allowance Trading Program. *Social Science Quarterly* 92(2): 297–323.

Roberts, D. 2020. "US Public Opinion Supports Action on Climate Change—and Has for Years." *Vox*. June 23. https://www.vox.com/energy-and-environment/2020/6/23/21298065/climate-clean-energy-public-opinion-poll-trends-pew.

Rolston, J. 2014. *Mining Coal and Undermining Gender: Rhythms of Work and Family in the American West*. New Brunswick, NJ, London: Rutgers University Press.

Roth, S. 2020. "Clean Energy Jobs Are Coming. Here's How to Make Sure They're Good Jobs." *Los Angeles Times*. September 3. https://www.latimes.com/environment/newsletter/2020-09-03/how-to-make-sure-clean-energy-jobs-are-good-jobs-boiling-point.

Ruttan, V.W. 2006. *Is War Necessary for Economic Growth?: Military Procurement and Technology Development*. Oxford, England: Oxford University Press.

Sanzillo, T. 2017. *A Transition Plan for Communities Affected by the Closings of Navajo Generating Station and Kayenta Mine*. Institute for Energy Economics and Financial Analysis. https://ieefa.org.

Schmalensee, R., and R.N. Stavins. 2017. Lessons learned from three decades of experience with cap and trade. *Review of Environmental Economics and Policy* 11(1): 59–79.

Schumpeter, J.A. 1934. *The Theory of Economic Development: An Inquiry into Profits, Capital, Credits, Interest, and the Business Cycle*. Piscataway, NJ: Transaction Publishers.

Shonkoff, S.B., R. Morello-Frosch, M. Pastor, and J. Sadd. 2011. The climate gap: Environmental health and equity implications of climate change and mitigation policies in California—A review of the literature. *Climatic Change* 109(1): 485–503.

Smil, V. 2010. *Energy Transitions: History, Requirements, Prospects*. Santa Barbara, CA: Praeger.

Somini, S., and V. Penney. 2020. "Big Tech Has a Big Climate Problem. Now, It's Being Forced to Clean Up." *The New York Times*. July 21. https://www.nytimes.com/2020/07/21/climate/apple-emissions-pledge.html.

Sovacool, B.K., D. Furszyfer Del Rio, and S. Griffiths. 2020. Contextualizing the COVID-19 pandemic for a carbon-constrained world: Insights for sustainability transitions, energy justice, and research methodology. *Energy Research and Social Science* 68.

State of California. 2020. "California Climate Investments." http://www.caclimateinvestments.ca.gov/cci-data-dashboard.

Steg, L., G. Perlaviciute, and E. van der Werff. 2015. Understanding the human dimensions of a sustainable energy transition. *Frontiers in Psychology* 6.

Storrow, B. 2020. "The Navajo, Circled by Coal, See Jobs Vanish as CO$_2$ Falls." *E&E News*. https://www.eenews.net/stories/1061970419.

Taylor, D.E. 2014. *The State of Diversity in Environmental Organizations*. Green 2.0 Working Group. Ann Arbor, MI: University of Michigan.

Thakrar, S.K., S. Balasubramanian, P.J. Adams, I.M. Azevedo, N.Z. Muller, S.N. Pandis, C.W. Tessum, et al. 2020. Reducing mortality from air pollution in the United States by targeting specific emission sources. *Environmental Science and Technology Letters* 7(9): 639-645.

Tierney, S.F., and P.J. Hibbard. 2002. Siting power plants in the new electric industry structure: Lessons from California and best practices for other states. *The Electricity Journal* 15(5): 35–50.

Tyson, A., and B. Kennedy. 2020. "Two-Thirds of Americans Think Government Should Do More on Climate." https://www.pewresearch.org/science/2020/06/23/two-thirds-of-americans-think-government-should-do-more-on-climate/.

UN FAO (United Nations, Food and Agriculture Organization). 2016. *Free Prior and Informed Consent: An Indigenous Peoples' Right and a Good Practice for Local Communities*. https://www.un.org/en/desa/products/publications.

Ungar, L., J. Barrett, S. Nadel, R. N. Elliott, E. Rightor, J. Amann, P. Huether, and M. Specian. 2020. "Growing a Greener Economy: Job and Climate Impacts from Energy Efficiency Investments." White paper. Washington DC: American Council for an Energy-Efficient Economy.

U.S. Bureau of Labor Statistics. 2020. "Occupational Outlook Handbook." https://www.bls.gov/ooh/fastest-growing.htm.

Van Slyke, A. 2020. "The Collapse of Health Care: The Effects of COVID-19 on US Community Health Centers." https://lernercenter.syr.edu/2020/08/10/ib-38/.

Vanclay, F. 2003. International principles for social impact assessment. *Impact Assessment and Project Appraisal* 21(1): 5–12.

Vetter, D. 2020. "EU's $1 Trillion Plan To Make Whole Continent Carbon Neutral." Forbes. https://www.forbes.com/sites/davidrvetter/2020/01/15/eus-1-trillion-plan-to-make-whole-continent-carbon-neutral/#118e29253c30.

Victor, D. 2019. "How Artificial Intelligence Will Affect the Future of Energy and Climate." Brookings Institution. http://brookings.edu/research/how-artificial-intelligence-will-affect-the-future-of-energy-and-climate/.

Walsh, K.B. 2017. Split estate and Wyoming's orphaned well crisis: The case of coalbed methane reclamation in the Powder River Basin, Wyoming. *Case Studies in the Environment* 1(1): 1-8.

Walsh, K.B., and J.H. Haggerty. 2018. "Governing Unconventional Legacies: Lessons from the Coalbed Methane Boom in Wyoming." Pp. 51–65 in *Governing Shale Gas* London: Routledge.

White-Newsome, J.L. 2016. A policy approach toward climate justice. *The Black Scholar* 46(3): 12–26.

Whitehouse, S. 2015. "The Fossil-Fuel Industry's Campaign to Mislead the American People." *Washington Post*. May 29.

Wiersma, B., and P. Devine-Wright. 2014. Public engagement with offshore renewable energy: A critical review. *Wiley Interdisciplinary Reviews: Climate Change* 5(4): 493–507.

Wilkinson, C. 2004. *Fire on the Plateau: Conflict and Endurance in the American Southwest*. Washington, DC: Island Press.

Williamson, V. 2018. "What France's Yellow Vest Protests Reveal about the Future of Climate Action." Brookings Institution. https://www.brookings.edu/blog/fixgov/2018/12/20/what-frances-yellow-vest-protests-reveal-about-the-future-of-climate-action/.

Wilsdon, J., and R. Willis. 2004. See-through science: Why public engagement needs to move upstream. *Demos* 1.

Wiser, R.H., and M. Bolinger. 2019. *2018 Wind Technologies Market Report*. Berkeley, CA: Lawrence Berkeley National Laboratory.

Wu, D. 2019. Accountability relations and market reform in China's electric power sector. *Global Transitions* 1: 171–180.

Wyborn, C., A. Datta, J. Montana, M. Ryan, P. Leith, B. Chaffin, C. Miller, and L. van Kerkhoff. 2019. Co-producing sustainability: Reordering the governance of science, policy, and practice. *Annual Review of Environment and Resources* 44: 319–346.

Zabin, C. 2020. *Release: Putting California on the High Road: A Jobs and Climate Action Plan for 2030*. Berkeley, CA: University of California Berkeley Labor Center.

CHAPTER FOUR

How to Achieve Deep Decarbonization

INTRODUCTION

This chapter addresses the policy package needed to achieve the first 10 years of the energy transition. These policies would accomplish the five quantitative technical objectives identified in Chapter 2 (targeting efficiency, electrification, zero-carbon power, infrastructure, and innovation), and place the nation on a 30-year path to net zero, while retaining optionality about the nature of the midcentury system. The package would also address the four societal goals developed in Chapter 3: enhanced U.S. economic leadership, an equitable transition and net-zero energy system, protected regional interests and sustained local communities, and cost-effectiveness.

To tackle both the technical and social needs, four overarching policy priorities are identified, each comprising a portfolio of specific proposals:

1. Establish the U.S.' **commitment** to a rapid, just, and equitable transition to a net-zero carbon economy.
2. Set **rules and standards** to accelerate the formation of markets for clean energy that work for all.
3. **Invest** in the research, technology, people, and infrastructure for a U.S. net-zero carbon future.
4. **Assist** families, businesses, communities, cities, and states in accelerating an equitable transition, ensuring that disadvantaged and at-risk communities do not suffer disproportionate burdens.

Table 4.1 summarizes how the specific policies support the technical and societal objectives. The combination of policies in this diverse portfolio shown in Table 4.1 is required to achieve all of the technical objectives while addressing multiple societal goals. There is no silver-bullet policy any more than there is a silver-bullet technology. Equally important, the policy portfolio in the table would greatly reduce climate disruption risks, increase long-term climate resilience, reduce air pollution and related health burdens, increase energy security, and support a clean energy industry and workforce.

TABLE 4.1 Summary of Policies Designed to Meet Net-Zero Carbon Emissions Goal and How the Policies Support the Technical and Societal Objectives

Policy	Technological Goals	Socioeconomic Goals	Government Entities	Appropriation, if Any	Notes
Establish U.S. commitment to a rapid, just, equitable transition to a net-zero carbon economy.					
U.S. CO$_2$ and other GHG emissions budget reaching net zero by 2050.			Executive and Congress	$5 million per year.	Budget is central for imposing emissions discipline, although any consequences for missing the target must be implemented through other policies. Funds are primarily for administration of the budget and data collection and management.
Economy-wide price on carbon.			Congress	None. Revenue of $40/tCO$_2$ rising 5% per year, which totals approximately $2 trillion from 2020 to 2030.	Carbon price level not designed to directly achieve net-zero emissions. Additional programs will be necessary to protect the competitiveness of import/export exposed businesses.
Establish 2-year federal National Transition Task Force to assess vulnerability of labor sectors and communities to the transition of the U.S. economy to carbon neutrality.			Congress	$5 million per year.	Task force responsible for design of an ongoing triennial national assessment on transition impacts and opportunities to be conducted by the Office of Equitable Energy Transitions.

Establish White House Office of Equitable Energy Transitions. • Establish criteria to ensure equitable and effective energy transition funding. • Sponsor external research to support development and evaluation of equity indicators and public engagement. • Report annually on energy equity indicators and triennially on transition impacts and opportunities.			Congressional appropriation	$25 million per year, rising to $100 million per year starting in 2025.	Federal office establishes targets and monitors and advances progress of federal programs aimed at a just transition.
Establish an independent National Transition Corporation to ensure coordination and funding in the areas of job losses, critical location infrastructure, and equitable access to economic opportunities and wealth, and to create public energy equity indicators.			Congressional appropriation	$20 billion in funding over 10 years.	Primary means to mediate harms that occur during transition, including support for communities that lose a critical employer, support for displaced workers, abandoned site remediation, and opportunities for communities to invest in a wide range of clean energy projects.

continued

165

TABLE 4.1 Continued

Policy	Technological Goals	Socioeconomic Goals	Government Entities	Appropriation, if Any	Notes
Set rules/standards to accelerate the formation of markets for clean energy that work for all.					
Set clean energy standard for electricity generation, designed to reach 75% zero-emissions electricity by 2030 and decline in emissions intensity to net-zero emissions by 2050.			Congress	None.	
Set national standards for light-, medium-, and heavy-duty zero-emissions vehicles, and extend and strengthen stringency of Corporate Average Fuel Economy (CAFE) standards. Light-duty zero-emission vehicle (ZEV) standard ramps to 50% of sales in 2030; medium- and heavy-duty to 30% of sales in 2030.			Congress	None.	
Set manufacturing standards for zero-emissions appliances, including hot water, cooking, and space heating. Department of Energy (DOE) continues to establish appliance minimum efficiency standards. Standard ramps down to achieve close to 100% all-electric in 2050.			Congress	None.	

Action			Authority	Funding	Notes
Enact three near-term actions on new and existing building energy efficiency, two by DOE/Environmental Protection Agency (EPA)[a] and one by the General Services Administration (GSA).			DOE, GSA	None.	GSA to set a cap on existing and new federal buildings that declines by 3% per year.
Enact five federal actions to advance clean electricity markets, and to improve their regulation, design, and functioning.[b]			Congress	$8 million per year for Federal Energy Regulatory Commission (FERC) Office of Public Participation and Consumer Advocacy.	Two of these actions involve FERC utilizing existing authorities and three involve congressional actions, two directed to FERC and one to DOE.
Deploy advanced electricity meters for the retail market, and support the ability of state regulators to review proposals for time/location-varying retail electricity prices.			Congressional appropriation for DOE	$4 billion over 10 years.	
Recipients of federal funds and their contractors must meet labor standards, including Davis-Bacon Act prevailing wage requirements; sign Project Labor Agreements (PLAs) where relevant; and negotiate Community Benefits (or Workforce) Agreements (CBAs) where relevant.			Congress	None.	

continued

TABLE 4.1 Continued

Policy	Technological Goals	Socioeconomic Goals	Government Entities	Appropriation, if Any	Notes
Report and assess financial and other risks associated with the net-zero transition and climate change by private companies, government agencies, and the Federal Reserve. Private companies receiving federal funds must also report their clean energy research and development (R&D) by category (wind, solar, etc.).			Congress	None.	Risk disclosures to be included in annual SEC reports for private companies. Federal Reserve to use climate-related risks in financial stress tests. Federal agencies to include climate-related risks in all benefit cost analyses. All banks to report on comparative financial investments in all energy sources.
Ensure that Buy America and Buy American provisions are applied and enforced for key materials and products in federally funded projects.			Congress	None.	
Establish an environmental product declaration library to create the accounting and reporting infrastructure to support the development of a comprehensive Buy Clean policy.			Congressional appropriation for EPA and DOE	$5 million per year.	

Invest (research, technology, people, and infrastructure) in a U.S. net-zero carbon future.				
Establish a federal Green Bank to finance low- or zero-carbon technology, business creation, and infrastructure.		Congressional authorization and appropriation	Capitalized with $30 billion, plus $3 billion per year until 2030.	Additional requirements include public reporting of both energy equity analyses of investment and leadership diversity of firms receiving funds.
Amend the Federal Power Act and Energy Policy Act by making changes to facilitate needed new transmission infrastructure.[c]		Congress	None.	
Plan, fund, permit, and build additional electrical transmission, including long-distance high-voltage, direct current (HVDC). Require fair public participation measures to ensure meaningful community input.[d]		Congressional authorization and appropriation for DOE and FERC	$25 million per year to DOE for planning; $50 million per year for DOE and FERC to facilitate use of existing rights-of-way; finance build through Green Bank; $10 million per year to DOE for distribution system innovations.	Funds provide support for technical assistance to states, communities, and tribes to enable meaningful participation in regional transmission planning and siting activities. Funds to distribution utilities to invest in automation and control technologies.

continued

TABLE 4.1 Continued

Policy	Technological Goals	Socioeconomic Goals	Government Entities	Appropriation, if Any	Notes
Expand electric vehicle (EV) charging network for interstate highway system.[e]			Congressional directive to Federal Highway Administration (FHWA) and National Institute of Standards and Technology (NIST); congressional appropriations to DOE	$5 billion over 10 years to expand charging infrastructure.	FHWA to expand its "alternative fuels corridor" program. NIST to develop interoperability standards for level 2 and fast chargers. DOE to fund expansion of interstate charging to support long-distance travel and make investments for EV charging for low-income businesses and residential areas.
Expand broadband for rural and low-income customers to support advanced metering.			Congress to authorize and fund rural electric cooperatives and private companies to offer broadband	$0.5 billion for rural electric cooperatives and $1.5 billion for private companies.	10% of investment costs to expand capabilities of smart grid to underserved areas. Grants or loans to rural electric providers and investment tax incentives to companies, both focused on rural and low-income communities.

170

Plan and assess the requirements for national CO$_2$ transport network, characterize geologic storage reservoirs, and establish permitting rules.[f] Require fair public participation measures to ensure meaningful community input.		Congressional authorization and appropriation to multiple agencies	$50 million to Department of Transportation (DOT) with other agencies involved for 5-year planning plus $50 million for block grants for community and stakeholder engagement. $10 billion to $15 billion total during the 2020s to DOE, United States Geological Survey (USGS), and Department of the Interior (DOI) to characterize reservoirs. Extend 45Q and increase to $70/tCO$_2$—$2 billion per year.	Modeling studies and other analysis indicate that significant amounts of negative emissions will be needed to meet net-zero emissions. The CO$_2$ pipeline network is needed even with 100% non-fossil electric power to enable carbon capture at cement and other industrial facilities with direct process emissions of greenhouse gases and to enable capture of CO$_2$ from biomass or via direct air capture for use in production of carbon-neutral liquid and gaseous fuels.

continued

TABLE 4.1 Continued

Policy	Technological Goals	Socioeconomic Goals	Government Entities	Appropriation, if Any	Notes
Establish educational and training programs to train the net-zero workforce, with reporting on diversity of participants and job placement success.[g]			Congressional appropriations to Department of Education, DOE, and National Science Foundation (NSF)	$5 billion per year for GI Bill-like program. $100 million per year for new undergraduate programs. $50 million per year for use-inspired and $375 million per year for other doctoral and postdoctoral fellowships. Eliminate visa restrictions for net-zero students. $7 million over 2020–2025 for the Energy Jobs Strategy Council.	Fields covered include science, engineering, policy, and social sciences, for students researching and innovating in low-carbon technologies, sustainable design, and the energy transition.
Revitalize clean energy manufacturing.[h]			Congressional appropriation and direction of Green Bank and U.S. Export-Import Bank	Manufacturing subsidies for low-carbon products starting at $1 billion per year and phased out over 10 years. No additional appropriation required for loans and loan guarantees from Green Bank and Export-Import Bank.	Export-Import Bank should make available at least $500 million per year in low-carbon product and clean-tech export financing and eliminate support for fossil technology exports.

172

Increase clean energy and net-zero transition RD&D that integrates equity indicators.[i]			Congressional appropriation for and directions to DOE and NSF	DOE clean energy RD&D triples from $6.8 billion per year to $20 billion per year over 10 years. DOE funds studies of policy evaluation at $25 million per year and regional innovation hubs at $10 million per year; DOE- and NSF-funded studies of social dimensions of the transition should be supported by an appropriation of $25 million per year.	Establish criteria for receiving funds on equity analysis, appropriate community input, and leadership diversity of companies applying for public investments. DOE to report on equity impacts and diversity of entities receiving public funds.
Increase funds for low-income households for energy expenses, home electrification, and weatherization.			Congressional appropriation	Increase Weatherization Assistance Program (WAP) funding to $1.2 billion per year from $305 million per year. Direct Department of Health and Human Services (HHS) to increase state's share of Low Income Home Energy Assistance Program (LIHEAP) funds for home electrification and efficiency.	

TABLE 4.1 Continued

Policy	Technological Goals	Socioeconomic Goals	Government Entities	Appropriation, if Any	Notes
Increase electrification of tribal lands		👥	Congressional appropriation to DOE and U.S. Department of Agriculture (USDA)	$20 million per year for assessment and planning through DOE Office of Indian Energy Policy (DOE-IE) and USDA Rural Utilities Service (USDA-RUS); expand DOE-IE to $200 million per year.	Increase direct financial assistance for the build-out of electricity infrastructure through DOE-IE grant programs.
Assist families, businesses, communities, cities, and states in an equitable transition, ensuring that the disadvantaged and at-risk do not suffer disproportionate burdens.					
Please note that the primary policies targeting fairness, diversity, and inclusion during the transition are the establishment of the Office of Equitable Energy Transitions and the National Transition Corporation, which are the fourth and fifth policies in this table.					
Establish National Laboratory support to subnational entities for planning and implementation of net-zero transition.	⚡ ⚙️ 🏠 💡 🗼	👥 🇺🇸	Congressional appropriation	Additional funding to national laboratories' annual funding commencing at the level of $200 million per year, rising to $500 million per year by 2025, and $1 billion per year by 2030.	To establish a coordinated, multi-laboratory capability to provide energy modeling, data, and analytic and technical support to cities, states, and regions to complete a just, equitable, effective, and rapid transition to net zero.

Action		Mechanism	Budget	Notes
Establish 10 regional centers to manage socioeconomic dimensions of the net-zero transition.ʲ		Congressional authorization and appropriations to DOE	$5 million per year for each center; $25 million per year for external research budget to provide data, models, and decision support to the region.	Coordinated by the Office of Equitable Energy Transitions.
Establish net-zero transition office in each state capital.		Congressional appropriations	$1 million per year in matching funds for each state.	Coordinate state's effort with federal and regional efforts.
Establish local community block grants for planning and to help identify especially at-risk communities. Greatly improve environmental justice (EJ) mapping and screening tool and reporting to guide investments.		Congressional appropriations to DOE	$1 billion per year in grants administered by regional centers.	Required to qualify for funding from the National Transition Corporation. Block grant funding requires inclusive participation and engagement by historically marginalized and low-income groups.

continued

TABLE 4.1 Continued

KEY TO ICONS
DARK GREEN icon indicates that the policy is highest priority and indispensable to achieve the objective. **MEDIUM GREEN** icon indicates that the policy is important to achieve the objective. **LIGHT GREEN** icon indicates that the policy would play a supporting role. No icon indicates that the policy would have at most a small positive role in achieving the objective (and might in, some cases, have a small negative impact on the objective).
Technological Goals
Invest in energy efficiency and productivity. Examples include accelerating the rate of increase of industrial energy productivity (dollars of economic output per energy consumed) from the historic 1% per year to 3% per year.
Electrify energy services in transportation, buildings, and industry. Examples include, by 2030, moving half of vehicle sales (all classes combined) to EVs, and deploying heat pumps in one-quarter of residences.
Produce carbon-free electricity. Roughly double the share of electricity generated by carbon-free sources from 37% to 75%.
Plan, permit, and build critical infrastructure. Build critical infrastructure needed for the transition to net zero, including new transmission lines, an EV charging station network, and a CO_2 pipeline network.
Expand the innovation toolkit. Triple federal support for net-zero RD&D.
Socioeconomic Goals
Strengthen the U.S. economy. Use the energy transition to accelerate U.S. innovation, reestablish U.S. manufacturing, increase the nation's global economic competitiveness, and increase the availability of high-quality jobs.
Promote equity and inclusion. Ensure equitable distribution of benefits, risks, and costs of the transition to net zero. Integrate historically marginalized groups into decision making by ensuring adherence to best-practice public participation laws. Require that entities receiving public funds report on leadership diversity to ensure nondiscrimination.
Support communities, businesses, and workers. Ensure support for those directly and adversely affected by the transition.
Maximize the cost-effectiveness of the transition to net zero.

[a] Direct DOE/EPA to expand their outreach of and support for adoption of benchmarking and transparency standards by state and local government through the expansion of Portfolio Manager. Direct DOE/EPA to further investigate the development of model carbon-neutral standards for new and existing buildings that, in turn, could be adopted by states and local authorities. Policies targeting retrofits of existing buildings will be in the final report.

[b] FERC should work with regional transmission organizations (RTOs) and independent system operators (ISOs) to ensure that markets in all parts of the country are designed to accommodate the shift to 100% clean electricity on the relevant timetable. Congress should clarify that the Federal Power Act does not limit the ability of states to use policies (e.g., long-term contracting with zero-carbon resources procured through market-based mechanisms) to support entry of zero-carbon resources into electric utility portfolios and wholesale power markets. Congress should further direct FERC to exercise its rate-making authority over wholesale prices in ways that accommodate state action to shape the timing and character of the transitions in their electric resource mixes. Congress should reauthorize the FERC Office of Public Participation and Consumer Advocacy to provide grants and other assistance to support greater public participation in FERC proceedings. FERC should direct North American Electric Reliability Corporation (NERC) to establish and implement standards to ensure that grid operators have sufficient flexible resources to maintain operational reliability of electric systems. Congress should direct and fund DOE to provide federal grants to support the deployment of advanced meters for retail electricity customers as well as the capabilities of state regulatory agencies and energy offices to review proposals for time/location-varying retail electricity prices, while also ensuring that low-income consumers have access to affordable basic electricity service.

[c] (1) Establish National Transmission Policy to rely on the high-voltage transmission system to support the nation's (and states') goals to achieve net-zero carbon emissions in the power sector. (2) Authorize and direct FERC to require transmission companies and regional transmission organizations to analyze and plan for economically attractive opportunities to build out the interstate electric system to connect regions that are rich in renewable resources with high-demand regions; this is in addition to the traditional planning goals of reliability and economic efficiency in the electric system. (3) Amend the Energy Policy Act of 2005 to assign to FERC the responsibility to designate any new National Interest Electric Transmission Corridors and to clarify that it is in the national interest for the United States to achieve net-zero climate goals as part of any such designations. (4) Authorize FERC to issue certificates of public need and convenience for interstate transmission lines (along the lines now in place for certification of gas pipelines), with clear direction to FERC that it should consider the location of renewable and other resources to support climate-mitigation objectives, as well as community impacts and state policies as part of the need determination (i.e., in addition to cost and reliability issues) and that FERC should broadly allocate the costs of transmission enhancements designed to expand regional energy systems in support of decarbonizing the electric system.

[d] (1) Congress should authorize and appropriate funding for DOE to provide support for technical assistance and planning grants to states, communities, and tribal nations to enable meaningful participation in regional transmission planning and siting activities. (2) Congress should authorize and appropriate funding for DOE and FERC to encourage and facilitate use of existing rights-of-way (e.g., railroad; roads and highways; electric transmission corridors) for expansion of electric transmission systems. (3) Congress should authorize and appropriate funding for DOE to analyze, plan for, and develop workable business model/regulatory structures, and provide financial incentives (through the Green Bank) for development of transmission systems to support development of offshore wind and for development, permitting, and construction of high-voltage transmission lines, including high-voltage direct-current lines.

TABLE 4.1 Continued

[e] (1) Congress should direct the Federal Highway Administration (a) to continue to expand its "alternative fuels corridor" program, which supports planning for EV charging infrastructure on the nation's interstate highways, and (b) to update its assessment of the ability and plans of the private sector to build out the EV charging infrastructure consistent with the pace of EV deployment needed for vehicle electrification anticipated for deep decarbonization, the need for vehicles on interstate highways and in public locations or high-density workplaces, and to identify gaps in funding and financial incentives as needed. In coordination with FHWA, DOE should provide funding for additional EV infrastructure that would cover gaps in interstate charging to support long-distance travel and make investments for EV charging for low-income businesses and residential areas. (2) NIST should develop communications and technology interoperability standards for all EV level 2 and fast charging infrastructure.

[f] Extend 45Q tax credit for carbon capture, use, and sequestration for projects that begin substantial construction prior to 2030 and make tax credit fully refundable for projects that commence construction prior to December 31, 2022. Set the 45Q subsidy rate for use equal to $35/tCO_2$ less whatever explicit carbon price is established and the subsidy rate for permanent sequestration to be equal to $70/tCO_2$ less whatever explicit carbon price is established. A hydrogen pipeline network will ultimately also be needed, but, as indicated in Chapter 2, the time pressure to build a national hydrogen pipeline network is less severe than for CO_2. This is because hydrogen production facilities can be located close to industrial hydrogen consumers, unlike CO_2 pipelines, which must terminate in geologic storage reservoirs. Also, hydrogen can be blended into natural gas and transported in existing gas pipelines, and gas pipelines could ultimately be converted to 100% hydrogen.

[g] (1) Congress should establish a 10-year GI Bill–type program for anyone who wants a vocational, undergraduate, or master's degree related to clean energy, energy efficiency, building electrification, sustainable design, or low-carbon technology. Such a program would ensure that the U.S. workforce transitions along the physical infrastructure of our energy, transportation, and economic systems. (2) Congress should support the creation of innovative new degree programs in community colleges and universities focused uniquely on the knowledge and skills necessary for a low-carbon economic and energy transformation. (3) Congress should provide funds to create interdisciplinary doctoral and postdoctoral training programs, similar to those funded by the National Institutes of Health (NIH), which place an emphasis on training students to pursue interdisciplinary, use-inspired research in collaboration with external stakeholders that can guide research and put it to use in improving practical actions to support decarbonization and energy justice. (4) Congress should provide support for doctoral and postdoctoral fellowships in science and engineering, policy, and social sciences for students researching and innovating in low-carbon technologies, sustainable design, and energy transitions, with at least 25 fellowships per state to ensure regional equity and build skills and knowledge throughout the United States. (5) The Department of Homeland Security (DHS) should eliminate or ease visa restrictions for international students who want to study climate change and clean energy at the undergraduate and graduate levels, where appropriate. (6) Congress should pass the Promoting American Energy Jobs Act of 2019 to reestablish the Energy Jobs Strategy Council under DOE, require energy and employment data collection and analysis, and provide a public report on energy and employment in the United States.

h (1) Congress should establish predictable and broad-based market-formation policies that create demand for low-carbon goods and services, improve access to finance, create performance-based manufacturing incentives, and promote exports. Specifically, Congress should provide manufacturing incentive through loans, loan guarantees, tax credits, grants, and other policy tools to firms that are matched with corresponding performance requirements. Subsidies provided directly to manufacturers must be tied to the meeting of performance metrics, such as production of products with lower embodied carbon or adoption of low-carbon technologies and approaches. Specific items could include expanding the scope of the energy audits in the DOE Better Plants program and expanded technical assistance to focus on energy use and GHG emissions reductions at the 1,500 largest carbon-emitting manufacturing plants; supporting the hiring of industrial plant energy managers by having DOE provide manufacturers with matching funds for 3 years to hire new plant energy managers; enabling the development of agile and resilient domestic supply chains through DOE research, technical assistance, and grants to assist manufacturing facilities in addressing supply chain disruptions resulting from COVID-19 and future crises. (2) Congress should provide loans and loan guarantees to manufacturers to produce low-carbon products, ideally through a Green Bank (see Chapter 4). (3) Congress should require the U.S. Export-Import Bank to phase out support for fossil fuels and make support for clean energy technologies a top priority with a minimum of $500 million per year. (4) Congress should create a new Assistant Secretary for Carbon Smart Manufacturing and Industry within DOE.

i (1) Congress should triple the DOE's investments in low- or zero-carbon RD&D over the next 10 years, in part by eliminating investments in fossil-fuel RD&D. These investments should include renewables, efficiency, storage, transmission and distribution (T&D), carbon capture, utilization, and storage (CCUS), advanced nuclear, and negative emissions technologies and increase the agency's funding of large-scale demonstration projects. By eliminating investments in non-carbon capture and storage (CCS) fossil-fuel RD&D, the net increase to the energy RD&D budget will be partially offset. (2) Congress should direct DOE to fund energy innovation policy evaluation studies to determine the extent to which policies implemented (both RD&D investment and market-formation policies) are working. (3) Congress should direct DOE and the NSF to create a joint program to fund studies of the social, economic, ethical, and organizational drivers, dynamics, and outcomes of the transition to a carbon-neutral economy, as well as studies of effective public engagement strategies for strengthening the U.S. social contract for decarbonization. (4) Congress should direct DOE to establish regional innovation hubs where they do not exist or are critically needed using funds appropriated under item 1 above. (5) Congress should direct DOE to enhance public-private partnerships for low-carbon energy.

j (1) Congress should coordinate federal agency actions at the regional scale through the deployment of federal agency staff to regional offices. (2) Congress should host a coordinating council of regional governors and mayors that meets annually to establish high-level policy goals for the transition. (3) Congress should establish mechanisms for ensuring the effective participation of low-income communities, communities of color, and other disadvantaged communities in regional dialogue and decision making about the transition to a carbon-neutral economy. (4) Congress should provide information annually to the White House Office of Equitable Energy Transitions detailing regional progress toward decarbonization goals and benchmarks for equity.

When introduced in Congress, most policies are assigned an "appropriations cost," which is an estimate of the dollar amount of federal appropriation required to implement the policy, if any. This cost is neither the capital investment required to achieve a technical objective, as reported in Chapter 2, nor is it the "social cost" used by economists to capture the consumption forgone by households and obtained from a general equilibrium economic model of the global economy.

Until recently, most of the legislative approaches proposed in Congress have been built around a single, overarching carbon pricing policy, such as cap-and-trade or a carbon tax (S. 2877, 111th Cong., 2009; Baker III et al., 2017). Broad carbon pricing policies are typically designed to satisfy an efficiency or cost-effectiveness test based on long-standing economic arguments (Hahn and Stavins, 1992). Under the assumption that the carbon price completely addresses climate-related externalities, other policies are justified only to the extent that they address other market failures, including information gaps, spillovers, other externalities, and market power (Jaffe et al., 2004, 2005; Driscoll et al., 2015; Newbery, 2008; Cohen, 1995). Equity and justice concerns, if addressed at all in a carbon pricing policy, have tended to be accomplished through the allocation of revenue from carbon taxes or auction of emissions allowances (S. 2877, 111th Cong., 2009; Baker III et al., 2017; RGGI, Inc., 2020; Green and Knittel, 2020).

While an economy-wide carbon price plays an important role in the presented policy roadmap, it does not do all the heavy lifting for several reasons. The existence of other market failures justifies a range of complementary interventions (Doris et al., 2009). These include federal emissions standards (e.g., Corporate Average Fuel Economy [CAFE]/greenhouse gas [GHG] emissions standards for light-duty vehicles); state standards and other state policies (e.g., California Zero Emissions Vehicle [CA ZEV] standards, Northeast (NE) states Regional Greenhouse Gas Initiative emissions allowance scheme); local standards (e.g., New York City Carbon Challenge); and corporate initiatives (e.g., Mars and WalMart's climate action plans). Equity and justice concerns are also placed alongside cost-effectiveness as equal if not more important goals. The direct effect of carbon pricing on gasoline, fuel oil, natural gas, and electricity is particularly regressive (Metcalf, 2008; Rausch et al., 2011; Williams et al., 2015; Green and Knittel, 2020), although this is true of other emissions-mitigation policies. Meanwhile, high carbon prices can affect competitiveness of U.S. industries exposed to international competition and trade (Aldy and Pizer, 2015). Carbon revenue allocation can attenuate these impacts at carbon prices up to $40 per ton, which is why carbon pricing proposals in the United States almost always include them. However, the committee is unaware of any studies examining whether this is possible at the higher prices necessary for deep decarbonization.

A number of recent approaches move the idea of carbon pricing to the side and focus directly on equity and justice through a larger set of more targeted policies (H. Res.109,

116th Cong., 2019; U.S. Congress, House, 2020a). One challenge in these proposals is to identify, mechanically, how all the pieces fit together to achieve the emissions goal.

The approach recommended here combines an overarching but insufficient (to achieve net-zero emissions) economy-wide carbon price and greenhouse gas budget with an additional set of policies that are all essential to address equity and justice considerations and to drive decarbonization in key sectors. The sector-by-sector approach presented here is consistent with emerging legislative text, which has pivoted from economy-wide solutions to sector-specific climate interventions (H.R.2486, 116th Cong., 2019; S.2300, 116th Cong., 2019; U.S. Congress, House, 2020a). Just as reaching net-zero emissions requires a full toolkit of low-carbon technologies (see Chapter 2), driving the net-zero transition requires the use of the full toolkit of policy levers. Also included is a mechanism to provide feedback if the policies need to be strengthened to meet the net-zero emission goal. The committee notes that this approach can be motivated either as a necessary deviation from cost-effectiveness and a heavy emphasis on carbon pricing (an economist framing) or as a logical consequence of addressing a fundamental system problem and transformation (Rosenbloom et al., 2020). Others have argued this approach is likely to promote public support (Bergquist et al., 2020; Brückman and Bernauer, 2020; Cullenward and Victor, 2020).

Beyond suggesting an extremely high carbon tax, there has been little research on a policy mix that can achieve net-zero emissions. The committee found no research on how to achieve the reductions needed as well as meet the diverse societal goals the committee lays out in Chapter 3. Rather than proposing more research to develop an "optimal" climate policy from the ground up, the committee has chosen to make recommendations that build on existing ideas where possible. This has the added benefit of stakeholder coalitions that have arisen around such proposals both in the United States and internationally. Moreover, the committee has sought to put them together to form a coherent pathway that puts the energy system on a trajectory to a net-zero economy by 2050.

This chapter explains the policy package needed to achieve net-zero emissions. Policies are organized thematically. The first set of policies, including a carbon price, is meant to establish the overall tenor and direction of the U.S. *commitment* to reducing GHG emissions through a small number of policies, including a carbon price and with equity, social justice, and engagement front and center. The second set defines the rules and regulations necessary to further align private incentives with overarching goals, including flexible, sector-specific zero-emissions performance *standards*. The third set of policies clarifies priorities for government *investment* along with incentives for private-sector investment. The last set rounds out additional policies necessary to *assist* in a fair and equitable transition to a net-zero emissions economy.

The following presents the rationale for each policy area, followed by evidence and implementation details related to each item within the area.

ESTABLISHING THE U.S. COMMITMENT TO A RAPID, JUST, AND EQUITABLE TRANSITION TO A NET-ZERO CARBON ECONOMY

This first policy domain emphasizes the policies that together establish the direction and tone of climate-change policy going forward. These policies include:

1. An economy-wide CO_2 and other greenhouse gas budget;
2. A price on carbon with appropriate measures to address competitiveness, equity, and environmental justice;
3. A framework and specific actions and commitments for justice and equity as integral elements of the low-carbon transition; and
4. A new social contract to connect public values to energy-system design.

A Greenhouse Gas Budget for the U.S. Economy

The starting point for decarbonization is to establish an overarching, economy-wide, cumulative GHG emission budget for the next several decades that produces an emissions trajectory leading to zero net emissions by midcentury. As discussed in prior chapters, net zero means that any remaining emissions at midcentury must be offset by negative emissions technologies such as afforestation, carbon capture and sequestration at electricity or industrial facilities, or direct ambient air capture and sequestration. (See Figure 2.2.)

For the United States, a net-zero target means that its net GHG emission budget between 2020 and 2050 is about 86 Gt CO_2e assuming a linear phase down from emissions of net 5.7 Gt CO_2e in 2020 to near zero in 2050.

A national emissions budget provides an unambiguous metric to assess whether policies are on track. The United Kingdom adopted a carbon budget in its Climate Change Act of 2009, where, in order to reach 80 percent emissions reduction by 2050, the government set up budgets for 5-year periods to serve as mileposts along the way to the 2050 target. The package of policies described in this report results in a robust suite of actions, incentives, investments, and transition-support programs, but these alone may not be enough, or their stringency may need to be tightened periodically. In particular, industrial emissions sources and a number of others, such as existing building equipment and nonroad transportation, face the economy-wide carbon price but are not otherwise directly regulated in our package. If, over time, the cumulative emissions budget is not achieved, these sectors may need direct regulation, the economy-wide carbon price may need to be raised, or zero-emission investment and/or technology incentives may need to be increased.

In this way, the policy provides the short- and medium-term price certainty of a carbon tax along with longer-term emissions certainty. That is, the budget provides a look-back mechanism as discussed in the referenced papers to make policy adjustments—including the tax level—depending on observed cumulative emissions. Unlike an ordinary cap-and-trade, which provides greater emission certainty and leaves cost uncertain, fixing a carbon price (through a carbon tax or a cap-and-trade with a price collar) leaves emissions uncertain (Weitzman, 1974; Burtraw et al., 2010; Fell et al., 2011). Therefore, other measures such as those discussed in the following sections may be needed to meet the cumulative emissions path and address adverse impacts on low-income communities and communities of color. Metcalf (2009) first proposed such measures. More recent discussions include two symposium discussions in the *Harvard Environmental Law Review* (2017) and *Review of Environmental Economics and Policy* (2020) (Aldy et al., 2017; Murray et al., 2017; Aldy, 2017; Hafstead et al., 2016; Brooks and Keohane, 2020; Aldy, 2020; Hafstead and Williams, 2020; Metcalf, 2020).

The committee recommends:

- Congress should enact a national, cumulative, greenhouse gas emission budget, similar to Figure 2.2, that goes to net-zero in 2050 and that establishes separate sectoral benchmarks for net CO_2 emissions from all sectors (industry, buildings, transportation, electricity, agricultural operations, net emissions from bio-energy with carbon capture and sequestration, and negative emissions from direct air capture, mineralization, forestry and agricultural soils, methane, nitrous oxide, and other non-CO_2 greenhouse gases). With critical funding for the mandate, the Environmental Protection Agency (EPA) should report annually on current and projected progress against the budget and for key technological benchmarks in the industry, buildings, transportation, and electricity sectors. For strategic action in the building sector, EPA's Portfolio Manager database that tracks measured energy use for U.S. buildings should enable prioritized actions for investing in building energy efficiency. Congress should further authorize and direct EPA to develop and report environmental indicators for areas where localized emissions and poverty pose environmental justice concerns.

    Cost: $5 million/year.

A Price on Carbon with Appropriate Measures to Address Competitiveness and Equity

As noted above, economy-wide carbon pricing is important to encourage emission reductions and to achieve net-zero emissions at the lowest cost. Carbon pricing is widely acknowledged by economists to be the key ingredient to achieve cost-effectiveness based on its ability to create consistent incentives throughout the economy to reduce emissions (Mufson, 2020). This is true along a pathway to zero emissions as well (Wigley et al., 1996).

But these same discussions also note that such a policy will need to include expenditures and programs, in particular, to avoid or mitigate inequities that will otherwise accompany such a policy, including impacts on low-income households and communities of color long exposed to the local air pollution that accompanies fossil-fuel combustion in power plants, buildings, vehicles, and industrial facilities. Additionally, carbon price policy should be designed in ways to avoid considerable disruption to trade flows in energy intensive industries highly exposed to import and export conditions.

The committee is not suggesting a carbon price do all the work, far from it. The regressive effects of carbon pricing on poor households is well documented (Metcalf, 2008; Rausch et al., 2011; Williams et al., 2015). Within income groups, Black households have higher residential energy expenditures than white households in the United States (Lyubich, 2020), so such a policy would have disproportionate effects on people of color. More generally, data show that even with very detailed socioeconomic information, there are considerable unexplained and irremediable differences in impacts across households (Pizer and Sexton, 2019; Rausch et al., 2011; Cronin et al., 2017; Fischer and Pizer, 2018; Green and Knittel, 2020). Thus, the typical response to addressing equity concerns with carbon pricing—directing payments to those adversely affected (Stavins, 2009)—only works to address broad regressivity or other easily targeted differences.

Distinct from equity concerns, there is the risk that carbon pricing will simply shift emissions and economic activity to jurisdictions with weaker regulation. This leads to both environmental (leakage) and economic (competitiveness) concerns (Jaffe et al., 1995; Frankel, 2008; Aldy and Pizer, 2015; Fischer and Fox, 2011). One way to address competitiveness is to design "border adjustments" for carbon pricing so that imports to the United States, and perhaps exports from the United States, are made competitive despite differences in carbon pricing. This is a complicated issue with distinct economic, political, legal, and practical issues (CBO, 2013; Kortum and Weisbach, 2017).

Another approach to address competitiveness impacts and some equity concerns has been to use carbon value to subsidize product prices (EPA, 2009; H.R. 2454,

111th Cong., 2009). That is, rather than giving revenue to those adversely affected by higher prices or foreign trade, revenue from carbon pricing is used to lower the price of emission/energy intensive industrial products facing trade competition and reduce electricity bill impacts. Electricity and industrial producers would still have the incentive to reduce emissions. End users, however, may lose their incentive to consume less.[1]

These discussions of equity and competitiveness concerns and the ability to ameliorate them hinge on the level of the carbon price itself. Most analyses and experiences concern relatively modest prices, ranging up to perhaps $30–40t/$CO_2$ (Cronin et al., 2017; EPA, 2009). The recent Climate Leadership Council proposes a price of $40t/$CO_2$ in 2021, rising at 5 percent per year (Climate Leadership Council, 2020). An exception is several recent carbon pricing proposals in the 116th Congress, some of which could reach between $75t/$CO_2$ by 2025 (C2ES, 2020a), which have not been analyzed for equity and trade impacts. Meanwhile, estimates of the price that would by itself drive to net-zero emissions by midcentury would be closer to at least $100t/$CO_2$ over the next decade and perhaps much higher in the future (Kaufman et al., 2020). Even these higher prices assume that certain "market failures" are addressed through complementary policies, including those that encourage electric vehicle adoption and improve vehicle fuel economy, and assumptions related to lower electricity demand, additional coal plant closures, and faster innovation (Kaufman et al., 2020). At these carbon prices, less is known about the effectiveness of policies to address equity and competitiveness concerns. It should be noted that the amount of revenue generated from prices of about $40t/$CO_2$ is approximately $2 trillion over a 10-year period (Horowitz et al., 2017; C2ES, 2019; Pomerleau and Asen, 2019). This revenue could be used for the funding of rebates and other activities to address the regressive nature of this policy and funding of clean energy investments.

With this in mind, the committee proposes *not* to select a carbon price designed to directly achieve net-zero emissions. Rather, it recommends that Congress adopt a policy meeting all of these objectives:

- Implement a carbon price of $40t/$CO_2$ in 2021 rising at 5 percent per year, targeting emissions from all uses of fossil fuels and industrial processes with GHG emissions. At these levels, existing research suggests equity and competitiveness concerns can be ameliorated. This would generate roughly $200 billion per year over the next decade, prior to any revenue use.

    Cost: Negative cost/positive revenue of approximately $200 billion/year.

---

[1] To address competitiveness and leakage, it is important to use allocation to lower product prices and encourage more domestic production. This contrasts with efforts to simply compensate affected industries.

- Address equity and fairness through both rebates and through funding of programs described in later sections within this chapter.
- Address competitiveness through a combination of output-based allocations and carbon border adjustments. These should target energy-intensive, trade-exposed industries. Output-based allocations should be designed to mitigate trade effects entirely, and carbon border adjustments should be implemented only if the output-based allocations prove insufficient. This may require additional research and data collection around the carbon dioxide embedded in traded goods and relevant carbon pricing along the value chain.

As described in the remainder of this chapter, this carbon price will then be combined with additional, harmonized companion policies to achieve net-zero emissions (Burtraw et al., 2018) in ways that address equity and competitiveness imperatives. While recognizing this *may* raise the overall monetary cost to society compared to an approach that uses carbon pricing as the primary tool to drive mitigation,[2] the committee's approach has the advantage of focusing on equity, fairness, and trade, as well as cost-effectiveness. Moreover, it is not clear from existing research whether the standard equity and competitiveness mechanisms will be effective under a pure carbon pricing approach designed to achieve net-zero emissions by 2050. At the same time, the proposed carbon price and companion policies do not alone ensure net-zero emissions. This requires the budget and look-back mechanism to raise the carbon price, strengthen existing policies, or enact additional policies in the future if cumulative emissions exceed the net-zero path.

An Equity and Social Justice Framework

As is clear from earlier in the report, the committee believes that however critically important and urgent it is to reduce GHG emissions, it must be done in ways that support a just and equitable transition. As discussed in Chapter 3, the costs and benefits of the current energy system are unequally distributed and create disproportionately negative impacts for disadvantaged populations, and, absent targeted policies and policy reform, this situation risks being repeated in a future energy system.

---

[2] To the extent the committee's companion policies simply correct other market failures (e.g., address innovation spillovers), they will not raise costs. However, the companion policies in the electricity, electric vehicle, and electric appliance market are designed to put an additional price on carbon emissions in these sectors. For example, one recent study suggests that a $150 per ton price might be needed to achieve 70 percent clean energy (implying an $85 per ton price for the clean energy standard). That same study also found only 41 percent of new car sales were electric vehicles even with a $150 per ton price (Larsen et al., 2020). Generally, this is not the most cost-effective way to address such additional market failures (Fischer and Newell, 2008).

An effective approach to address equity and social justice dimensions in national energy policy requires oversight and coordination, the establishment of key criteria and programs to monitor them, and mandated commitments to seek out and provide resources to enable and assimilate the perspectives of historically marginalized stakeholders and groups into energy system design. Running across these threads is the imperative to develop strategies that are both top-down and bottom-up—for example, by coordinating the development of tools and processes for vulnerability assessment at a national scale while meaningfully including local stakeholders in the deployment of such tools and interpretation of their findings.

The following federal actions are necessary to build and implement an equity and social justice framework as part of the energy transition:

- Congressional authorization of and appropriations for the convening of a 2-year National Transition Task Force comprised of nongovernmental community and expert stakeholders, with a directive for the Task Force to report to the White House Office of Equitable Energy Transitions, Congress, and the public on:
    - The vulnerabilities of U.S. labor sectors and communities to the transition of the U.S. economy to carbon neutrality;
    - The needs of diverse communities experiencing transition impacts and inequitable energy burdens, as well as research priorities to address these needs and the design of standards for an equitable and just transition;
    - A draft Presidential Policy Directive that would require relevant federal agencies to integrate equitable energy transition objectives into agencies' policies, programs, procurement decisions, project reviews, grants, and other administrative decisions, and to do so on an expedited and cooperative basis while also ensuring inclusion of meaningful participation by relevant agency staff (no additional cost);
    - The history of successes and failures in prior U.S. efforts to support distressed communities and regions facing diverse economic challenges, lessons to be learned for efforts to address equity concerns in decarbonization policy, and strategies for integrating equity responses with wider U.S. efforts to address inequality in society as a whole; the provision of greater opportunities for the labor force and stakeholders in vulnerable communities to derive value from the energy, including through enabling them to have access to investments in low-/no-carbon infrastructures and buildings;
    - The adequacy of existing federal programs and support for vulnerable communities affected by the energy transitions (e.g., those related to abandoned-mine lands, coal ash sites, brownfields redevelopment

programs), as well as recommendations for any needed changes in those programs or for any new programs to support equitable and just outcomes;
  ○ Analysis, insights, and recommendations to the EPA with regard to establishing cumulative air-emission limits, targets for local emissions reductions, and other environmental improvements (e.g., water quality, exposure to hazardous wastes) specific to local environmental justice communities;
  ○ Barriers and opportunities to successful and equitable public engagement processes for the planning of low-carbon energy systems;
  ○ Social, public health, and environmental risks of infrastructure abandonment from bankruptcies;
  ○ Federal decommissioning and remediation regulations, and the policy reforms needed, focusing on retired and retiring fossil-fueled generating plants and abandoned oil wells, natural gas wells, and coal mines while recommending a time frame to expand the analysis to other fossil fuel infrastructure; and
  ○ The design of a federal program for an ongoing triennial national assessment on transition impacts and opportunities with attention to the equity dimensions described above, with that assessment to be conducted by the Office of Equitable Energy Transitions (described below).

  Cost: $5 million/year for Transition Task Force.

- Congressional authorization of and appropriations for the establishment of an independent Office of Equitable Energy Transitions within the Executive Office of the President responsible for interagency coordination and assessment, analysis, and evaluation of the nation's energy transitions. The functions of this Office would be to:
  ○ Establish criteria to ensure equitable and effective allocation of energy transition funding;
  ○ Establish targets for key indicators, annually evaluate progress toward those goals, and conduct the triennial national assessment on transition impacts and opportunities;
  ○ Ensure that appropriate equity standards and assessments are incorporated into implementation of all federal energy and environmental programs and regulatory decisions;
  ○ Assess and make recommendations to rectify the lack of representation of affected groups and stakeholders on the Secretary of Energy's Advisory Board and other federal advisory committees (e.g., DOE's Electricity Advisory Committee and EPA's Clean Air Act Advisory Committee);

- Oversee and coordinate federal agencies' implementation of programmatic reforms in response to the public engagement evaluation conducted by the Task Force; and
- Sponsor external research to support its work in establishing equity criteria, developing and assessing key targets for tracking equity and effectiveness indicators, and implementing improvements in federal agencies' public engagement on the energy transition.

Cost: The annual cost to staff and fund the research, reporting, assessment, and other responsibilities of the Office of Equitable Energy Transitions will begin at $25 million per year, rising to $100 million/year starting in 2025.

A New Social Contract to Mitigate Harm and Expand Economic Opportunities for Impacted Communities

Chapter 3 recommends that any sustainable decarbonization strategy must build on a strong new social contract that commits to innovative and novel forms of public engagement and new pathways for creating public value from energy transitions. As indicated in Table 3.3.1 a wide range of communities either currently struggle or expect to be struggling with the impacts of climate change and of the energy transitions in diverse and sometimes multiple ways.

Building a social contract depends on ensuring equitable access to wealth generated by the transition, mitigating harms to vulnerable populations and geographies, pursuing new approaches to include diverse American voices in designing and creating low-carbon energy futures, and realigning how the public realizes value from and contributes to value in national energy policies and investments. Policy must also address socioeconomic and racial inequalities resulting from energy system architectures.

To these ends, the committee recommends that Congress:

- Establish a new federally chartered, independent National Transition Corporation (NTC) to complement the functions of the White House Office of Equitable Energy Transitions, to ensure coordination and funding in the areas of job losses, critical infrastructure, and equitable access to economic opportunities and wealth creation. The NTC would be tasked with the following objectives:
    - Coordinate and leverage existing federal programs and agencies to deliver employment, housing, small business assistance, and other critical social services through temporary initiatives focused on decarbonization impacts and opportunities;

- Deliver funding and implementation support for reclamation and remediation in the case of gaps caused by bankruptcies and asset orphaning;
- Provide opportunities for low-income communities to develop projects that ensure low-income communities have a direct stake in the clean energy transition;
- Demonstrate local commitment and provide direct distributions to replace critical public revenue shortfalls—including debt maintenance—based on eligibility and credit criteria; and
- Effectively engage diverse, broad-based stakeholder groups in oversight and implementation of NTC programs.

• The NTC would also have the responsibility to:
- Recommend changes to laws or regulations to expand the notification requirements and thresholds in the Worker Adjustment and Retraining Act to give vulnerable communities and labor sectors adequate time to plan for and secure resources for retraining;
- Establish an Energy Transition Jobs Initiative as a joint effort of the National Transition Task Force and Office of Equitable Energy Transitions, to aggregate and streamline delivery of support packages to transition frontline workers. This can be accomplished by updating the triggers and qualifying standards of economic adjustment programs to recognize the unique circumstances of transition frontline workers and to enable proactive planning and by extending support beyond the coal industry to extraction, processing, and distribution of other carbon-intensive energy resources;
- Fund major community-based demonstration projects that strengthen equity outcomes and further NTC objectives to support activities such as fund reclamation and remediation in the case of orphaned infrastructure and unfavorable bankruptcy proceedings; fund the implementation and enforcement of existing laws to accomplish reclamation and remediation; direct distributions to replace critical revenue shortfalls; fund development opportunities for low-income communities to invest in a wide range of clean energy projects, including distributed renewable energy, energy storage, microgrids, and transportation.

The value of a federally chartered corporation model is that it can be endowed with dedicated funding and empowered to act strategically in the interest of its charter, giving it the necessary autonomy to act both quickly and continuously. Stable sources of funding that can be used for relevant governmental purposes are essential to provide predictability and secure success of transition initiatives. The NTC will be governed by five members who are Senate-confirmed presidential appointees, with staggered 4-year terms and with no more than three members of the same party. That Board will select and hire a chief executive officer who reports directly to a Board of Directors. The members of the Board shall have relevant experience in working with economic development, communities in transition, persistent poverty geographies, and Black, Indigenous, people of color (BIPOC) communities.

The NTC will provide funding in the form of grants and other direct distributions to provide subsidies for certain private investments. The NTC will be directed to establish a formula to distribute the transition funds directly to local governments. The NTC formula should also include a cost share requirement for recipients. The NTC's distribution formula will prioritize locations currently experiencing an acute fiscal crisis associated with the actual or expected loss of revenue resulting from the closure of energy-generating or energy-refining facilities or from the decline or closure of resource extraction activities (e.g., coal, oil, and natural gas). Eligibility will also consider community characteristics including social and economic measures of income, poverty, education, geographic isolation, and others identified by the White House Office of Equitable Energy Transitions in the interest of identifying cases of past energy injustices.

Cost: $20 billion in funding over 10 years. This is based on $3 billion for the Energy Transition Jobs Initiative, up to $2 billion for reclamation work, and $15 billion to support communities through grants, loans, loan guarantees, and/or subsidies for development projects and direct distributions. Congress should provide an initial no-year appropriation (which can be held until it is used up) of $10 billion at the outset, with $1 billion a year in additional annual funding.

SETTING RULES AND STANDARDS TO ACCELERATE THE FORMATION OF MARKETS FOR CLEAN ENERGY THAT WORK FOR ALL

Because the carbon price recommended in this report will not be sufficient to drive decarbonization to net zero, specific sets of rules and standards are needed to guide private-sector decisions so that they are aligned with achieving decarbonization while realizing social goals. The first of these is a zero-emission standard for the power sector (also known as a clean energy standard). Others include energy-efficiency standards for appliances; energy efficiency standards for new and existing buildings; CAFE/GHG emissions standards for vehicle fleets; standards for the design of zero-carbon electricity markets; standards for labor engaged in clean-energy work; standards for corporate reporting of climate risk; and standards for U.S. government procurement.

The committee will address additional standards for other sectors of the economy—rail and air transportation, industrial energy use, and existing (versus new) buildings and vehicles—in its final report.

A Clean Energy Standard for Electricity

A clean energy standard for electricity is a relatively cost-effective way to eliminate emissions in the power sector that also mitigates some equity and competitiveness concerns. Simple carbon-pricing raises electricity prices for two reasons: the technological cost of producing electricity with less CO_2 (Palmer et al., 2018; Larson et al., 2018), and charges for the remaining CO_2 emissions (Fischer and Pizer, 2018). This "carbon charge" is a rent or payment that accrues to someone in the form of allowance value (if allowances are freely allocated under cap-and-trade) or to the government (if allowances are auctioned or under a carbon tax). It is generally paid by end users of electricity, and serves as an appropriate incentive to conserve electricity in order to reduce emissions further (Ho et al., 2008).

Many policies, proposed and implemented, suggest ways to use carbon revenue to depress adverse effects on electricity end users, including equity and competitiveness effects (California Climate Investments, 2020; H.R. 2454, 111th Cong., 2009; Tierney and Hibbard, 2019). Other carbon pricing programs in the electric sector—for example, the multistate Regional Greenhouse Gas Initiative—auction the allowances and then reinvest the proceeds in consumers' bill reductions or energy efficiency measures (which further reduce consumers' electricity bills; see Hibbard and Tierney, 2011; Hibbard et al., 2018). Others propose to give allowances to local utilities, who are instructed to use the allowance value to protect end-users (e.g., H.R. 2454, 111th Cong., 2009).

There is conflicting evidence if California's cap-and-trade program has yielded improvements in environmental equity with respect to health-damaging co-pollutant emissions. Cushing et al. (2018) presents evidence from California's cap-and-trade program showing emissions of co-pollutants associated with ambient air quality and human health effects (particulate matter, nitrogen oxides, sulfur oxides, volatile organic compounds, and air toxics) increasing in socioeconomically disadvantaged communities. However, in a recent study, Hernandez-Cortes and Meng (2020) suggest that the program has reduced the pollution exposure gap between disadvantaged and other communities.

An alternative approach is a clean energy standard (CES) in the power sector (Aldy, 2011). Such a policy addresses some equity and competitiveness concerns by depressing the price effects on end users relative to simple carbon pricing. This policy still involves the potential for certain justice concerns, particularly if credit trading leads to more emissions in disadvantaged communities. Moreover, the committee still recommends additional competitiveness and especially equity-related policies elsewhere in this report. In one design, allowances would still be required for GHG emissions from electric generators, as under carbon pricing. Allocation, however, would be based on the volume of electricity generation and the established standard "performance rate" (this is sometimes called a rate-based approach). Individual generators are typically credited or debited based on their performance relative to the standard.[3] Generators buy and sell credits in a market, which establishes a transparent price. A policy to achieve carbon neutrality in the power sector would gradually ramp the performance rate to zero. For example, with the U.S. power sector currently emitting roughly 0.45 tons of CO_2 per megawatt hour (EIA, 2020), a policy that started with a performance rate of 0.45 and declined to zero by 2050, would fully decarbonize the power sector. There are a number of additional design options and nuances in this type of policy that are discussed in the literature (e.g., Aldy, 2011; C2ES and RAP, 2011; Fitzpatrick et al., 2018; Cleary et al., 2019).

A second approach to the design of a CES would focus on requiring sellers of retail electricity to rely on an increasing share of zero-carbon sources. This approach would operate along the lines of the current renewable portfolio standards (RPS) that have been adopted by 30 states and the District of Columbia, or like the CES adopted in 4 states (DSIRE, 2019). Under a similarly designed national CES, the policy could call for increasing amounts of zero-carbon supply, expressed as a percent or share of total sales, with a target year for reaching a 100 percent. Each year, retail sellers of electricity need to demonstrate that they have a power supply portfolio that satisfies the required percentage of zero-carbon resources. Retail sellers with excess zero-carbon generation can sell credits to sellers with deficits, such that the overall national system hits the target. This approach would help to pull zero-carbon resources into the system while increasingly restricting fossil generation that does not include carbon capture (Cleary et al., 2019; U.S. Congress, House, 2020b). It is generally criticized, however, in not discriminating among higher and lower emitting fossil fuel sources on the pathway to zero emissions (Aldy, 2011).

It should be noted that Congress has recently introduced multiple CES bills (S.1359, 116th Cong., 2019; S.1974, 116th Cong., 2019; H.R.7516, 116th Cong., 2020) and the House Climate Crisis Committee Report released in June 2020 also featured CES, indicating existing political support and momentum for this approach.

---

[3] For example, if a coal plant emits 1 ton per megawatt-hour as it produces 100 megawatt hours, and the standard is 0.2 tons/MWh, it will owe the regulator (1 ton/MWh—0.2 tons/MWh) × 100 MWh = 80 tons worth of credit. Meanwhile, low carbon electric generators, including zero-emitting sources, earn credits based on the amount they beat the standard and the amount of electricity that they sell. A zero emitting source facing the same 0.2 tons/MWh standard, and generated 100 megawatt hours, would earn 20 credits (denominated in tons of carbon dioxide).

Note that a power-sector standard policy would operate alongside the economy-wide carbon price that would also cover the electricity sector. To the extent that the economy-wide price is sufficient to decarbonize the power sector, the CES will have little effect. However, it is anticipated that the chosen economy-wide price will not be sufficient. The CES will provide the necessary additional incentives to drive the sector to zero emissions.

The committee recommends that Congress:

- Adopt a clean energy standard for electricity along the lines of Aldy (2011) designed to reach roughly 75 percent clean electricity share by 2030 and a declining emissions intensity reaching zero net emissions in 2050.

Electrification and Efficiency Standards for Vehicles, Appliances, and Buildings

As noted earlier in this report, reaching a net-zero economy will require significantly and rapidly reducing power sector emissions and the electrification of a substantial portion of vehicles, buildings, and appliances. Moreover, it is critical to pursue substantially increased energy efficiency in order to reduce the total amount of electric capacity needed to meet demand and to help control energy costs. The overarching carbon-pricing policy described in the earlier section will likely be insufficient to drive demand reduction as a critical step in effective low-carbon electrification.

Minimum energy efficiency standards for appliances, building efficiency standards, and average vehicle fuel-economy standards have been long used to drive increased energy efficiency and energy productivity (Alliance Commission, 2013; Nadel et al., 2015). There is a long-running debate in economics about the role of these types of standards, and whether decisions regarding the purchase of energy efficient equipment are subject to various market failures (Hausman and Joskow, 1982; Fischer, 2004; Jaffe et al., 2004; Gillingham et al., 2004, 2006; Houde and Spurlock, 2016). There has also been discussion of shifting the minimum standards for appliances to average standards, similar to those for vehicles, to increase cost-effectiveness (USG, 2017). There have been pro and con arguments for such changes, with some asserting that the added flexibility would reduce compliance costs for manufacturers and prices for consumers and others arguing it would add undue levels of complexity to program administration and allow standards to backslide (Blonz et al., 2018; Urbanek, 2017). Nonetheless, there is general recognition that these standards have been shown to drive increased efficiency (Doris et al., 2009), to avoid fuel consumption, and to reduce GHG emissions (Greene et al., 2020).

Energy use in buildings accounts for approximately 28 percent of total U.S. energy consumption, taking into account both buildings' direct use of energy and their use of electricity (DOE, 2020a). One approach that has been shown to help achieve efficiency improvements is to measure a building's energy use, benchmark it (e.g., relative to its own past use or to comparable buildings or to an advanced "stretch" building code), and then provide the information to the building's owner, manager or occupant (EPA, 2012; Palmer and Walls, 2015; Meng et al., 2016). Such benchmarking helps to drive the market for efficiency services and reduction in buildings' energy use. Policies relating to building codes and standards have typically been the domain of states and local government. The DOE has supported policy assessments and the provision of information to stakeholders, but even so, as of January 2020, only 35 U.S. localities and 3 states had adopted benchmarking and transparency policies that require reporting of energy consumption for public and privately owned commercial and/or multifamily buildings (IMT, 2020). The federal government should expand its outreach of and support for adoption of benchmarking and transparency standards by state and local government.

Given its status as the largest landlord in the United States (Jungclaus et al., 2017), the federal government also has a more direct role to play in making its buildings more energy efficient and less carbon intensive. The federal government should set an emissions cap for existing and new federally owned buildings, with the cap declining at 3 percent per year (Architecture 2030, 2014) and with emissions reductions accomplished through energy efficiency upgrades, switching to electric or district systems, and/or generating/procuring carbon-free renewable energy. These federal-building emissions caps would be models for states and municipalities to set standards for buildings with public and private sector ownership (such as has already occurred in New York City, whose 2019 Local Law No 97 requires large existing buildings to reduce their emissions by 40 percent by 2030 and 80 percent by 2050, from a 2005 baseline; see NYC, 2019).

In order to drive further energy efficiency for appliances, buildings, and transportation, existing programs and policies will need to be adapted and strengthened in the future. Existing laws allow DOE to set appliance standards to levels "technologically feasible and economically justified," but regulatory action has varied over time (Clark, 2019). Vehicle standards are focused on increasing miles-per-gallon and reducing emissions for gasoline and diesel vehicles, not improving efficiency of future electric vehicles (C2ES, 2020b). Resources exist for states to continue to improve building regulations (DOE, 2020b; California, 2020a). Further work is necessary to strengthen these standards in preparation for increases in electrification.

There is less experience with direct electrification regulation itself. California has mandated a certain fraction of passenger vehicle sales to be zero-emission vehicles (ZEVs) since 1998, although this standard was modified frequently in its early stages as vehicle batteries lagged in their development (NRC, 2006; Collantes and Sperling, 2008).

The current requirement will reach 22 percent by 2025 (California, 2020b) and California's governor has recently issued an executive order requiring sales of all new passenger vehicles to be zero emission by 2035 (Office of the Governor, 2020). Ten other states (Colorado, Connecticut, Maine, Maryland, Massachusetts, New Jersey, New York, Oregon, Rhode Island, and Vermont) have adopted California's ZEV requirements for their own vehicle fleets (EDF, 2019). EV policies could be extended and expanded to require increased vehicle electrification at the national level for the ground transportation sector. Policies should expand the current focus beyond light-duty vehicles, to include medium- and heavy-duty vehicles.

To further electrify household and commercial appliances (heating, hot water, and cooking) will require additional policies. Appliance electrification policies could mirror the EV mandates, requiring manufacturers to sell an increasing fraction of electric products with the flexibility to trade among manufacturers. Alternatively, the policies could be focused on emissions per product to be reduced over time to zero, similar to the clean energy standard for electricity described above.

Distinct from increasing restrictions on fossil fuel equipment, a number of jurisdictions have recently adopted building codes to encourage electrification. This includes policies to reduce access to natural gas (Margolies, 2020) or require all-electric appliances through "reach" codes (DiChristopher, 2020). The California Energy Commission is preparing a modification to its Building Energy Efficiency Standards to mandate new construction to be all electric starting in 2023. The Rocky Mountain Institute found that delaying an all-electric construction requirement to the 2025 code cycle would result in 3 million additional tons of carbon emissions by 2030 and more than $1 billion of spending on new gas infrastructure (Grab and Shah, 2020). Building codes to drive electrification could be encouraged at the federal level but would be implemented at the state level in the United States, given state-level authority (Vaughan and Turner, 2013).

The committee recommends that Congress:

- Direct EPA to establish a national zero-emission vehicle standards. They should be set on a timetable to achieve 50 percent of new sales of light-duty vehicles and 30 percent of sales of medium- and heavy-duty vehicles by 2030 (either EVs or fuel-cell vehicles).
- Direct EPA/DOT to continue tightening light-duty vehicle fuel economy/greenhouse gas emissions standards beyond model year 2026.
- Direct DOE to establish a national zero-emission appliance manufacturing standard covering all fossil-emitting building uses (space heating/cooling, hot water, and cooking). This should be modeled after the ZEV vehicle standards and achieve full electrification by 2050.

- Reaffirm that the DOE continue to establish minimum efficiency standards for appliances, particularly targeting electric appliances.
- Direct DOE/EPA to expand their outreach of and support for adoption of benchmarking and transparency standards by state and local government through the expansion of Portfolio Manager.
- Direct DOE/EPA to further investigate the development of model carbon neutral standards for new and existing buildings that, in turn, could be adopted by states and local authorities.
- Direct the General Services Administration (GSA) to set an aggregate emissions cap for existing and new federal buildings, with the cap declining at 3 percent per year. GSA should prioritize high-reduction, low-cost actions.

Cost: None of these actions would require an additional appropriation by Congress beyond the program management resources.

Improved Regulation and Design of Power Markets for Clean Electricity

Given the outsized role that the electric sector will need to play in a low-carbon energy economy, electric systems need to operate efficiently and reliably, to attract capital for significant new infrastructure investment in a timely way, and to provide economically accessible power for all Americans. In conjunction with the overarching market-based policies to explicitly price and directly drive down power-sector CO_2 emissions to net zero, the structure and design of retail utility regulation and wholesale electricity markets together need to support such investment, operations, and reliability. Wholesale market design, combined with state and federal policies, will play key roles in enabling new zero-carbon resources to enter the market as rapidly as possible (and for others to remain in operation, where current power market conditions do not support continued operations of certain existing zero-carbon resources in the absence of carbon prices).

It is well understood that tomorrow's electric system will depend increasingly on low-carbon resources with high upfront capital costs and very low operating costs (Aggarwal et al., 2019; Bielen et al., 2017; Corneli, 2018; Ela et al., 2014; Pierpont and Nelson, 2017.) This is a different set of conditions than those in place when many regions of the United States adopted centrally organized energy and capacity markets for electric power (Clements, 2017; Joskow and Schmalensee, 2020). Even with a national policy that prices carbon emissions into electricity markets and requires the share of zero-emission generation to rise to 100 percent, conditions in the future will tend to produce very low electric-energy prices during more and more hours of the year. In turn, revenues in wholesale energy markets alone are not likely to be sufficient to support accelerated entry (and maintenance) of zero-carbon technologies in many regions of the county.

Distinct from designing wholesale markets that work with increasing and eventually 100 percent zero-carbon sources, many states are interested in pursuing their own efforts. The federal government should encourage rather than discourage those efforts. Wholesale markets will need to allow for states' policy-driven market-based instruments (such as competitive power procurements leading to long-term contracts for off-shore wind, storage, carbon capture, utilization and storage, and other technologies), and this may require Congress to direct the Federal Energy Regulatory Commission (FERC) to accommodate such state-supported approaches. (This might be akin to the provisions of the amended Federal Power Act that state that "no wholesale transmission order may be issued that is inconsistent with any state law governing retail marketing areas of electric utilities" [16 U.S.C. 824k(g)], which was intended to harmonize states' decisions regarding the structure of the electric industry in their states with FERC's role in encouraging open access to transmission.)

Although today's wholesale market designs vary across these regions, all of the Regional Transmission Organizations (RTOs)/Independent System Operators (ISOs) that operate the markets use bid-based markets for wholesale electricity with security-constrained economic dispatch and locational-marginal pricing mechanisms. Such markets are the gold standard for efficient operations of a portfolio of resources in place at any point in time (Fox-Penner, 2020; Joskow and Schmalensee, 2020; Hogan, 2014, 2017). Some argue that energy-only wholesale markets (e.g., without capacity markets) with opportunity-cost pricing and bilateral contracting will perform well in the future (Hogan, 2017; Gramlich and Hogan, 2019). Stakeholders in many regions of the United States, however, may not support such an approach. Analysis also suggests that such designs are not likely to support entry of clean energy resources on a fast-enough time frame consistent with the nation's decarbonization needs (Fox-Penner, 2020; Joskow and Schmalensee, 2020). Also, it is not clear that the centralized capacity markets in several RTOs are sustainable as they are currently configured, because there is so much tension in states' efforts to support contracts that retain or pull zero-carbon resources into the market.

Those parts of the United States with traditional utilities and no retail competition may be better positioned for investment in zero-carbon technologies in light of rate-base treatment of fixed costs and the ability for utilities to sign long-term contracts with third-party suppliers (Joskow and Schmalensee, 2020; Corneli et al., 2019; Fox-Penner, 2020.) In these markets, states already can use mechanisms such as least-cost planning, competitive power procurements, and utility investments to shape their supply portfolios. (Many of the committee's recommendations for federal action aim at encouraging these and other states to take more aggressive action to reduce carbon emissions from their power sector and elsewhere in local economies.)

In all parts of the United States, most electricity consumers will need to be exposed to real-time, locational pricing in order to provide flexible demand and to avoid the large capacity additions that would otherwise be needed in its absence. FERC has supported adoption of market rules in RTO wholesale markets to accommodate supply from distributed energy resources (FERC Order 2222; [FERC, 2020]), including specifically addressing energy storage (FERC Order 841; [FERC, 2018]) and demand response (FERC Order 745; [FERC, 2011]). Presumably at the retail level, there will be utilities and third-party intermediaries to provide different pricing and service-delivery options to consumers, but the former will need to be able to see real-time pricing. In parallel, there will need to be advanced meters to open up access to flexible demand and demand management strategies. The federal American Recovery and Reinvestment Act (ARRA) stimulus package provided approximately $3.4 billion to accelerate electric utilities' deployment of advanced meters and related infrastructure, and led to roughly 16 million meters being installed around the United States (DOE, 2015, 2016). As of 2018, however, nearly half of the nation's electricity meters—43 percent of residential meters, 46 percent of commercial meters, and 49 percent of industrial meters—did not have advanced two-way communications capability enabling visibility on real-time prices and supporting flexible demand (EIA, 2019).

The committee recommends that:

- FERC work with RTO/ISOs to ensure that markets in all parts of the country are designed to accommodate the shift to 100 percent clean electricity on the relevant timetable.
- Congress clarify that the Federal Power Act does not limit the ability of states to use policies (e.g., long-term contracting with zero-carbon resources procured through market-based mechanisms) to support entry of zero-carbon resources into electric utility portfolios and wholesale power markets. Congress should further direct FERC to exercise its rate-making authority over wholesale prices in ways that accommodate state action to shape the timing and character of the transitions in their electric resource mixes.
- Congress reauthorize FERC's Office of Public Participation and Consumer Advocacy to provide grants and other assistance to support greater public participation in FERC proceedings.

    Cost: $8 million/year.

- FERC direct the North American Electric Reliability Corporation (NERC) to establish and implement standards to ensure that grid operators have sufficient flexible resources to maintain operational reliability of electric systems.
- Congress direct and fund the Department of Energy to provide federal grants to support the deployment of advanced meters for retail electricity customers as

well as the capabilities of state regulatory agencies and energy offices to review proposals for time/location-varying retail electricity prices while also assuring that low-income consumers have access to affordable basic electricity service.

Cost: $4 billion over 10 years.

Labor Standards for Clean Energy Work

The transition to clean energy presents enormous opportunities for job growth in clean energy sectors, which is already occurring. In 2019, there were 3.6 million workers in clean energy jobs in the United States, including energy efficiency, electric and alternative fuel vehicles, solar energy, wind energy, biofuels, and battery storage (NASEO and EFI, 2019).

Clean energy jobs have higher wages than the national average and tend to have lower educational requirements, making them more accessible (Muro et al., 2019). However, the reality is that the energy transition thus far has largely displaced good-paying, stable, and high-benefits jobs and has not created jobs with comparable wages, benefits, locations, and hours (see Partridge and Steigauf, 2020, for example). An illustrative 2015 analysis of the Clean Power Plan, which would have mandated emissions reductions in existing power plants, showed that while net jobs were created, the jobs lost were less likely to be low wage and less likely to require a 4-year degree (Bivens, 2015). As stationary fossil fuel plants are retired and replaced by distributed wind and solar, this imbalance between the quality of jobs lost and the jobs gained can be mitigated with federal assistance, complementary policies, and the cooperation of organized labor. Additionally, ensuring that jobs created in clean energy are high-wage, safe, family-supporting jobs that enable communities and workers to capture the benefits of clean energy will maintain the social contract.

To ensure that such jobs are created in the transition, labor standards should be attached to federal funding and support for clean energy projects. The Davis-Bacon Act may be referenced as an existing standard that has an accepted framework for its use. The use of Department of Transportation and Department of Housing and Urban Development funds and their pass-through programs such as the Community Development Block all require compliance with the Davis-Bacon Act. Bids for utility-scale wind and solar development projects, which contain these types of policies, are already cost-competitive in many areas (such as California, for example), and good wages and benefits lead to a safer and more productive construction workforce that is highly skilled and trained (Jones et al., 2016). Even if labor standards increase the cost of labor, the cost of labor for installation of utility-scale wind and solar projects is less than 10 percent of the total development costs (Fu et al., 2018; Stehly and Beiter, 2019).

There are a number of pathways to increase wages. Historically, labor unions have been a pathway to the middle class and economic prosperity for Americans and a way to improve workers' wages (Voos, 2009; Ahlquist, 2017; Bivens et al., 2017; Farber et al., 2018). Although politically contentious, they have also proven to increase worker safety and reduce income inequality. Other ways to increase income include earned income tax credits and a minimum wage. For the past several decades, however, American workers have faced wage stagnation, rising income inequality, and coordinated efforts to remove their right to organize (Horowitz et al., 2020; Shierholz, 2019). To maintain the social contract for a transition to net zero, workers must be assured that the clean energy economy can work for them and that their rights will be protected.

The committee recommends that:

- Federal grants, loans, tax incentives, and other support for projects should be conditioned on recipients and their contractors meeting strong labor standards (including Davis-Bacon Act prevailing wage requirements, compliance with all labor, safety, environmental, and civil rights statutes), requiring that federally funded construction and infrastructure project developers sign Project Labor Agreements (PLAs) where relevant, and requiring recipients of federal incentives negotiate Community Benefits (or Workforce) Agreements (CBAs), where relevant.

   Cost: No direct additional costs to federal government.

Standards for Corporate Reporting

The financial performance of countless and quite-different American companies—which account for 88 percent of U.S. economic activity[4]—and the interests of both shareholders and workers will be affected by climate change. Many firms' assets, operations, and/or supply chains will be physically and financially impacted by a changing climate (e.g., from extreme weather events and temperature change). Others' business models are vulnerable to reputational risk or market competition. Many businesses will grow in a transition to a low-carbon economy. Others will be challenged because their operations and those of their suppliers face the possibility that public policy or litigation will require deep reductions in GHG emissions in the future. This is true for companies that are directly involved in the energy industries as well as companies in the larger economy whose businesses will be affected by incremental and fundamental changes in energy markets.

---

[4] This metric reflects 2018 value added by private industries as a percentage of gross domestic product (BEA, 2018).

Although the magnitude, timing, location, character, distribution, and costs of such climate-related risks (and opportunities) are uncertain (Weitzman, 2009), they are systemic and may lead to significant disruptions in markets, financial institutions, the economy, communities, and workers (Ramani, 2020).

Investors depend on well-functioning financial markets with transparent information. "Open economies of sound macroeconomic policies, good legal systems, and shareholder protection attract capital and therefore have larger financial markets" (World Bank, 2020).

Financial markets play an essential role in the economy by pricing risk "to support informed, efficient capital-allocation decisions," but many companies do not provide sufficient information to show that they adequately factor in climate-related risks. "More effective, clear, and consistent climate-related disclosure is needed from companies around the world" (TCFD, 2017).

Many financial risk-management experts observe that climate risk still is poorly priced into financial markets, in part because there is inadequate transparency in corporate financial statements and because it is difficult to assign probabilities on government action (Litterman, 2020a,b). Even recognizing growing investor interest in companies with positive environmental, social, and governance (ESG) practices and outcomes (Fink, 2020; Eccles and Klimenko, 2019), many companies have not integrated climate risk into their governance and fiduciary responsibilities (Zaidi, 2020).

Many investors, financial fiduciaries and other fund managers, and others have called for reforms in financial markets to address and internalize climate risk into companies' information disclosures (Vizcarra, 2020), and in their internal financial, economic and risk analyses, systems, metrics (TFCD, 2017). Several bills have been introduced in Congress to accomplish such objectives, and the House Select Committee on the Climate Crisis has recommended several legislative actions to "expose climate-related risks to private capital to shift assets toward climate-smart investments" (U.S. Congress, House, 2020b).

The committee recommends that Congress:

- Direct the Securities and Exchange Commission (SEC) to require public companies to formally disclose their risks from adverse impacts of climate change mitigation policies and climate change as part of their annual filings to the SEC.
- Direct the Federal Reserve to identify climate-related financial risks, including by applying climate change policy and impact scenarios to financial stress tests.

- Direct federal agencies (e.g., EPA, Department of Energy [DOE], Department of Transportation [DOT], House and Urban Development [HUD], FERC, SEC) to incorporate risks and costs from climate policies and climate change into the benefit-cost analyses required prior to the adoption of regulations or standards, or approval of public or private infrastructure investments).
- Require private firms to report their energy-related research and development investments by category (e.g., fossil, solar, wind) annually to the Department of Energy.

Cost: No cost beyond administrative.

In additional the committee recommends:

- The Commodities Future Trading Commission should build on the recommendations of the report *Managing Climate Risk in the U.S. Financial System* (Climate-Related Market Risk Subcommittee, 2020) to ensure that climate risk is better reflected in the commission's and other federal financial agencies' oversight of commodities and derivative markets.

U.S. Government Procurement Policy and Domestic Clean Energy Markets

Even with increasing deployment of clean technology, the U.S.' ability to manufacture such technologies is not keeping pace. In some instances, the United States depends on imports from other countries for materials and components critical to a clean economy. "Under current government procurement policies and trade rules, much of the public spending for infrastructure and clean energy systems would leak away to foreign providers, in the form of increased imports" (Scott, 2020).

Failure to produce these technologies domestically puts the United States at risk and threatens future jobs and the economy. Making these products in the United States is critical to leadership in the clean economy and necessary for innovation and global competitiveness. Developing solutions to the economics and foreign competition conundrum is an important part of developing a domestic clean energy market. However, while the United States needs to be able to produce final products like wind turbines and solar panels domestically, the majority of manufacturing jobs in many energy-related sectors are at supplier companies, not the end assembler or original equipment manufacturers. In the auto industry, for example, three out of every four manufacturing jobs are parts workers (Ruckelshaus and Leberstein, 2014). A robust domestic supply chain for these products is critical for innovation but also for resilience and to withstand disruption, which has become evident during the COVID-19 global pandemic.

"Buy American" or "Buy America" provisions require that projects funded directly or indirectly with federal dollars use specified products such as iron and steel made in the United States, ensuring that the United States maintains the ability to produce critical materials and products (Morgan, 2019). These provisions have been added to federal infrastructure bills and passed with bipartisan support: the bipartisan American Water Infrastructure Act (AWIA) of 2018 was passed with a Buy America provision requiring that drinking water infrastructure supported by funds from the Drinking Water State Revolving Fund is built with U.S.-made iron and steel (see American Iron and Steel provision in CRS, 2018). The ARRA of 2009 included a Buy American provision that required domestic sourcing of iron, steel, and manufactured goods for projects funded by the stimulus (DOE, n.d.).

Many industrial materials such as iron, steel, chemicals, cement, and concrete have high levels of embodied carbon emissions (see, e.g., Fischedick et al., 2014). To meet the goal of net-zero emissions by 2050, embodied carbon emissions in materials must decrease. A Buy Clean procurement policy will drive down embodied carbon emissions within products by establishing a baseline level of emissions intensity for key input materials and requiring that a percentage of materials procured achieve that baseline or lower. Focusing on federal, state, and local government procurements—which, according to expert testimony, account for the purchase of 90 percent of the cement and concrete and 50 percent of the steel used in the United States (Friedmann, 2019)—could create significant demand for cleaner materials and create a high-achievers market. Further, investments in innovation in materials and assemblies that reduce embodied carbon, including the development of alternative high-performance products that can be manufactured in the United States, could be achieved through dedicated National Science Foundation and DOE programs.

Deep decarbonization also means that the United States should have policies that help to avoid the leakage of emissions overseas, which occurs when the U.S. imports materials with high embodied carbon emissions. A recent report estimates that 25 percent of the world's total emissions pass through a carbon accounting loophole by not including embodied carbon emissions of imported products in the consuming country (Moran et al., 2018). While these emissions are being debited at the producer side, it can allow countries that import products with high embodied carbon emissions, such as steel and cement, to avoid fully accounting for this portion of their carbon footprint. A Buy Clean procurement policy would reduce the offshoring of U.S. emissions while strengthening clean U.S. manufacturing and increasing global competitiveness of U.S. industry.

Developing a Buy Clean standard will require a number of elements: deciding products for which Buy Clean applies; defining a standardized life cycle emissions accounting system (such as Environmental Product Declarations [EPDs]) so that emission intensity can be compared for those products; and setting a maximum emission intensity for each product. This accounting system should build on existing certification programs such as Energy Star. Stakeholder engagement with industry, academia, workers, and community groups to determine the products and materials covered and set the benchmarks should be undertaken to ensure a transparent decision-making process.

The State of California passed the Buy Clean California Act in 2017, which covers concrete-steel rebar, flat glass, structural steel, and mineral-wool board insulation and uses EPDs for emission intensity reporting. The Department of General Services is tasked with establishing the maximum emission intensity for products by January 2021 (CA DGS, 2018). The CLEAN Futures Act introduced in January 2020 would establish a similar Buy Clean program nationally (U.S. Congress, House, 2020a,c).

A comprehensive Buy Clean policy might include an additional requirement that a portion of procurements meet higher emissions standards, creating a high achievers' market to drive down emission intensity and cost. It would likely also include direct support for manufacturers to conduct life cycle analysis and report emission intensity of their products (the CLEAN Futures Act includes technical assistance for this) as well as make efficiency and technology improvements to lower their emissions. A "Buy Fair" component added to a Buy Clean standard would ensure that labor standards are met as well.

Establishing comprehensive policy and generating a set of standards will require a stakeholder engagement process and development of accounting and reporting infrastructure. An initial, immediate step is to begin to build the accounting and reporting infrastructure.

The committee recommends that Congress:

- Ensure that Buy American and Buy America provisions are appropriately applied and enforced to cover key materials and products on federally funded projects.

  Cost: No direct cost.

- Direct EPA and DOE to establish an EPD library to create the accounting and reporting infrastructure to support the development of a comprehensive Buy Clean policy.

  Cost: $5 million/year for EPA and DOE to cover information requirements and administrative needs.

INVESTING IN A NET-ZERO U.S. ENERGY FUTURE

Policies aimed at unleashing public and private investment will be required during the first 10 years of the energy transition. The necessary investments take many forms: investment in long-distance transmission of renewable energy or in EV-charging networks; investment in education and training to build a talented workforce that is fit for service in a low-carbon economy; investment in domestic manufacturing of clean energy technologies; investment in R&D for technology innovation and deployment; investment in understanding and mitigating the impacts of decarbonization on communities; and investment in building resilient communities in a low-carbon economy. The committee thus proposes a number of institutions and policy instruments designed to mobilize public and private investment in and financing of the energy transition.

Creation of a Green Bank

Although the transition might be achieved while spending only a fraction of gross domestic product (GDP) that the nation currently allocates to its energy system, the transition will be much more capital intensive than business-as-usual (Chapter 2). Private sources are unlikely to provide the needed capital, especially during the 2020s when the effort is new. To ensure industrial competitiveness and quality of life, the United States should establish a Green Bank to mobilize finance for low-carbon infrastructure and business in America. Partial financing by a Green Bank would reduce risk for private investors and encourage rapid expansion of private source capital. Such a bank would underpin the broad economic and social transitions required to achieve net-zero emissions by midcentury. The new bank should lend, provide loan guarantees, make equity investments, cooperate with community banks to increase the availability of finance at the local level, and leverage private finance consistent with a national strategy to compete internationally in low-carbon industries and transform the U.S. economy. It should make particular effort be a source of credit for innovative small and medium-size enterprises that may be locked out of commercial markets owing to their size. The Green Bank can be a lead investor on big decarbonization projects that serve the public good, de-risking and leveraging larger commercial investors. It should address inequities in the financing system, working with local banks, co-ops, and rural and other marginalized communities. It can also play a countercyclical role by scaling up lending operations when private banks contract (Luna-Martinez and Vicente, 2012), which is essential to sustained and uninterrupted access to finance during the low-carbon transition.

U.S. companies have to compete globally with German, British, Indian, and Chinese firms, among others, all supported by government-backed financial institutions that have a specific public policy mandate. The German KfW, UK Green Investment Bank, China Development Bank, and Industrial Development Bank of India are a few examples. The German KfW is one of the largest development banks in the world, with assets exceeding €500 billion. It was initially the sole lender in Germany to solar companies, prior to financing from private banks. The China Development Bank holds assets exceeding $1 trillion and likewise has invested heavily in renewable energy and low-carbon infrastructure (Griffith-Jones and Ocampo, 2018). The UK established the world's first green investment bank in 2012, which financed more than £12 billion of UK green infrastructure projects between 2012 and 2017. This bank backed the construction of the Rampion offshore wind farm and invested in four other offshore wind farms. In 2017, the UK government privatized the bank in order to access additional capital and pay off public debt. It was acquired by an Australian firm, Macquarie, and it now operates as the Green Investment Group. All of the taxpayer money was returned with a gain of £186 million, but the UK government announced in 2020 that it would create a new state-backed Green Bank in the UK.

The United States currently has no domestic independent development, investment, or Green Bank at the federal level, but it has periodically used them in the past. The War Finance Corporation was established during World War I to mobilize finance for the war effort, and in 1932, President Hoover created the Reconstruction Finance Corporation, which later became the capital bank for the New Deal (Omarova, 2020). However, federal agencies including DOE and U.S. Department of Agriculture (USDA) do have substantial programs to invest in domestic development through loans and loan guarantees, research grants, and loan and grant assistance. At the USDA for example, the Rural Energy for America Program administered by the Rural Business and Cooperative Service offers loans and grants to rural businesses and agriculture producers to adopt renewable and energy efficiency measures in their farm operations. At the subnational level, at least nine states have established Green Banks or funds, ranging from the Connecticut Green Bank to the Colorado Clean Energy Fund. There are also a number of local funds that serve specific communities, such as the Solar and Energy Loan Fund (SELF) in Florida. These investments also mobilize private sector investment into a project by reportedly three to six times the amount of public sector dollars at work (NREL, 2017). Legislation has been introduced into Congress for a National Climate Bank with an initial capitalization of $10 billion and an additional $5 billion per year for 5 years to reach $35 billion. The Coalition for Green Capital (2019) suggests this could mobilize up to $1 trillion in investment.

While an initial multi-billion-dollar capitalization for the Green Bank would be a significant investment of federal resources, it should be financially self-sustaining and assets should grow over time. There is no magic number for initial capitalization, but to enable the green recovery that is needed in the United States, it needs to be large enough to be adequate to the task and to compete with its counterparts. The China Development Bank's current assets equal $1 trillion, Germany's KfW's are $575 billion, and Brazil's National Development Bank is worth $145 billion. A recent proposal for an American Development Bank called for an initial capitalization of $100 billion (Griffith-Jones, 2020). The recent establishment of the U.S. Development Finance Corporation came with authorization of $60 billion, so an initial capitalization of $30 billion in a U.S. Green Bank, rising to $60 billion, may be politically realistic. Equal authorizations would establish that the government cares just as much about domestic investments in green economic development as it does in overseas investments.

The committee recommends that a federal Green Bank be established with a specific public mission to finance low- or zero-carbon technology, business creation, and infrastructure. The rationale for an independent Green Bank as opposed to an entity like a Clean Energy Deployment Administration is to allow it to operate more nimbly than would be the case if the Green Bank was a federal entity. An independent Green Bank formed by the federal government and capitalized with federal funds could forgive loans, something that most governmental entities cannot do. Its remit could be broader, encompassing the financing of other green industries and sectors (e.g., climate adaptation and resilience, fresh water supply), but it must devote at least two-thirds of its financing for the energy transition to achieve net-zero emissions by midcentury. Its objectives within the energy transition space would include fostering long-term domestic manufacturing capacity in clean energy and energy efficiency.

The committee recommends:

- Establishment of a federal Green Bank with a specific public mission to finance low- or zero-carbon buildings and technologies, business creation, and infrastructure.
- Congress should provide an initial capitalization of a minimum of $30 billion, followed by an additional $3 billion per year through 2030, resulting in a minimum capitalization of $60 billion by 2030.

  Cost: $60 billion.

- The bank must adopt good governance procedures and practices, including being transparent and abiding by environment and social safeguards and incorporating labor standards (and Buy American) requirements.
- The staff of the bank must be trained not only in finance but also in engineering, science, technology, and policy so that the bank can make well-informed investment decisions.

- The bank must devote at least two-thirds of its financing to the social, economic, and infrastructural energy transition to achieve net-zero emissions by midcentury.
- The bank must report annually to Congress on its investments and their impacts, including total financing, firms supported, infrastructure created, jobs created, value added, and reduced or avoided GHG emissions.

Invest in New Infrastructure

Like today's energy systems, a net-zero energy economy will require numerous energy-delivery systems and networks to connect energy sources with energy consumers. Some of these systems—like the high-voltage electric grid—will build on the current interconnected interstate transmission network. Others—such as an expansive body of EV charging stations that are as accessible as today's gasoline filling stations—will need to be developed from the relatively nascent stage that exists today. This policy cluster involves recommendations related to electric transmission, EV charging, deployment of broadband to underserved areas, and CO_2 pipelines. The committee's final report will discuss other infrastructure needs for the later decades, including transport of hydrogen.

Electric Transmission and Distribution Infrastructure

A net-zero energy economy that depends on both a decarbonized electric system and electrification of many building, vehicle, and industrial energy uses will require expansion of today's high-voltage electric grid and local distribution-system infrastructure. Even assuming significant deployment of distributed energy resources (e.g., solar panels, microgrids, energy efficiency, and flexible demand), the nation will also need an expanded high-voltage grid to connect regions with high-quality renewables to locations where people live and work (U.S. Congress, House, 2020b; MacDonald et al., 2016). The distribution system will need to be expanded to accommodate greater capacity requirements associated with electric vehicles, heat pumps, and distributed energy resources. It will also require investment in expanded automation and controls to handle more complicated power flows and to enable such things as greater demand response of EV charging and space and water heating loads, as well as cooling energy storage for air conditioning buildings.

With regard to the bulk power system, two persistent conditions threaten to undermine the ability of the country to scale up access to and development of high-quality renewables: First, a chicken-and-egg problem currently exists with respect to the development of high-quality renewable projects in remote areas (e.g., offshore wind, wind in the Prairie states) and access to transmission to ensure that that renewable power can be delivered to distant load centers. Second, the current federal/state jurisdictional split, in which FERC regulates transmission planning/access and the states determine whether to approve transmission facilities, has proven to stand in the way of building out the kind of high-voltage transmission system needed for deployment of renewables at scale (NASEM, 2017b; Reed et al., 2019). The approach approved by Congress in 2005 to designate National Interest Electric Transmission Corridors proved unsuccessful (Swanstrom and Jolivert, 2009; CRS, 2010).

An enhanced interstate transmission grid will require long-term national and regional electric-system planning. The current planning paradigm—for example, long-term transmission planning conducted by regional grid operators and transmission companies under FERC authority; DOE's analysis of congested transmission corridors; separation of planning for generation from planning for transmission in many if not most parts of the country—is not up to the task of what is needed to open up large regional markets for development of high-quality renewable resources. In the large portions of the country with RTOs/ISOs, such planning is designed to inform decisions of market participants on various potential wires/generation/demand-side solutions. While designed to support efficient outcomes, these approaches are insufficient to put in place, in a timely fashion, the kind of high-voltage interstate transmission system that is needed for deep decarbonization.

Planning for and siting of transmission requires many improvements: a national statement of the important role of transmission in supporting the nation's, regions', and states' achievement of GHG-emission reduction targets (U.S. Congress, House, 2020b); provision of "side-payments" or other economic incentives for states that need to host transmission enhancements for national and regional purposes (Reed et al., 2020; Eto, 2016); greater use of existing rights of way to site new transmission (Reed et al., 2020, 2019); financial support for state and local governments to analyze transmission projects and to provide meaningful analyses of barriers to local economic development through transmission, such as poorly designed incentive schemes (Haggerty et al., 2014); and support for authentic engagement of stakeholders, with community groups supported by resources so that they can meaningfully participate in regional planning processes (Eto, 2016). In the upcoming section on strengthening the capacity to effectively and equitably transition to a clean energy future, the committee recommends various policies and actions to support participation in regional energy/transmission plans.

With regard to the local distribution system, the committee anticipates that electric utilities will make customer-funded investments over time in response to and in anticipation of changes in demand and power flows on the local system. The committee believes, however, that the needed acceleration of electrification of building end uses and vehicles, combined with continuing requirements for reliable and affordable electricity supply, also warrants the availability of near-term federal incentives for investment in automation and control technologies on distribution systems.

The committee recommends that Congress:

- Amend the Federal Power Act to:
  - Establish a U.S. National Transmission Policy to enable a high-voltage transmission system to support the nation's (and states') goals to achieve net-zero carbon emissions in the power sector.
  - Authorize and direct FERC to require transmission companies and regional transmission organizations to analyze and plan for economically attractive opportunities to build out the interstate electric system to connect regions that are rich in renewable resources with high-demand regions; this is in addition to the traditional planning goals of reliability and economic efficiency in the electric system.
- Amend the Energy Policy Act of 2005 to assign to FERC the responsibility to designate any new National Interest Electric Transmission Corridors and to clarify that it is in the national interest for the U.S. to achieve net-zero climate goals as part of any such designations.
- Authorize FERC to issue certificates of public need and convenience for interstate transmission lines (along the lines now in place for certification of gas pipelines), with clear direction to FERC that it should consider the location of renewable and other resources to support climate-mitigation objectives, as well as community impacts and state policies as part of the need determination (i.e., in addition to cost and reliability issues) and that FERC should broadly allocate the costs of transmission enhancements designed to expand regional energy systems in support of decarbonizing the electric system.
- Authorize and direct FERC to approve compensation to states and tribes to compensate for lands traversed by existing and new transmission projects that support regional clean energy objectives.
- Authorize and appropriate funding for:
  - DOE to provide support for technical assistance and planning grants to states, communities, and tribes to enable meaningful participation in regional transmission planning and siting activities.

Cost: $25 million/year.

- DOE and FERC to encourage and facilitate use of existing rights-of-way (e.g., railroad; roads and highways; electric transmission corridors) for expansion of electric transmission systems.
- DOE to analyze, plan for, develop workable business model/regulatory structures, and provide financial incentives (through the Green Bank) for development of transmission systems to support development of offshore wind and for development, permitting, and construction of high-voltage transmission lines, including high-voltage direct-current lines.

Cost: $50 million/year for analysis and planning, and for technical assistance to states, tribes, localities. No incremental cost for the transmission lines (included in Green Bank).

- DOE to provide grants to local distribution utilities for innovative projects to encourage investment in automation and control technologies on distribution systems.

Cost: $10 million/year.

Electric Vehicle Charging Infrastructure

Decarbonizing the nation's energy economy will depend on rapid electrification of the vehicle fleet, which will, in turn, require the build-out of electric vehicle (EV) charging infrastructure.

Americans have come to expect that refueling their vehicles is convenient, given the near ubiquitous nature of the fuel-filling infrastructure. Today's filling stations are typically available within relatively close distances to homes, offices, and major thoroughfares, and the act of filling up a tank with gasoline or diesel fuel takes little time. Drivers' willingness to purchase and depend on EVs for their mobility needs depends upon their expectations that they will be able to charge their vehicles conveniently and relatively quickly. Broad adoption of EVs will be frustrated if consumers and workers lack access to EV charging infrastructure—whether at home, in parking lots, at office buildings, at local service stations, and at stops on interstate highways. Less than half of U.S. households have access to off-street parking and adequate electric service (Traut, 2013).

Planning for EV charging infrastructure has been undertaken in various localities and regions of the country, and the federal government and governors in many regions are cooperating with efforts to coordinate such planning on interstate routes (e.g., FHWA, 2020). Many private companies have invested in commercial charging facilities, and states have used a variety of approaches (e.g., use of the Volkswagen settlement funds; tax incentives) to create incentives for infrastructure development.

The National Governors Association reports that "many states are exploring the role of their electric utilities in building the EV charging network needed. State public utility commissions have already approved roughly $1 billion in utility EV infrastructure investments, with another $1.5 billion in additional utility investments already proposed" (NGA, 2019). In some states (e.g., Minnesota), where utilities have exclusive franchises to sell electricity to consumers, legislatures and utility regulators have established carve-outs where third parties may own EV charging stations that sell power to vehicle operators.

In spite of considerable work under way to support development of EV charging infrastructure, significant gaps may exist between the scope of EV charging infrastructure that is on the ground or on drawing boards, and the vast network of EV charging stations that will be needed to provide consumer confidence. In Chapter 2, the committee identified the goals of (1) 60 million light-duty EVs and trucks and 1 million medium-duty and heavy-duty vehicles, including buses, to be on the roads by 2030; and (2) 3 million public Level 2 charging units and 120,000 DC fast-charging units. As of May 2019, there were an estimated 58,000 Level 2 and 10,800 DC charging units throughout the United States (DOE, 2019). The Breakthrough Institute estimated the need for up to 9.6 million EV chargers by 2030 and calls for a federal investment of $5 billion (Olson, 2020). Like an earlier National Academies report (TRB and NRC, 2015) on barriers to electric vehicles, the Breakthrough Institute highlighted fast charging on interstate highway corridors as a particular area for investment.

Much more planning and investment for EV infrastructure development is needed, by the public and private sectors. Fleet operators could be leaders in this effort. To spur EV deployment and use, the federal government and states should accelerate planning and deepen financial incentives for EV charging infrastructure build-out. Particular attention must be paid to how future building designs and community planning accommodates access to convenient EV charging. Creating the opportunity for home-based charging to the roughly half of U.S. households that do not have a garage or at-home off-street parking will be essential. Also, the federal government should work with stakeholders to establish interoperability standards for the EV Level 2 and fast-charging infrastructure.

The committee recommends that Congress direct:

- The FHWA to
    - Continue to expand its "alternative fuels corridor" program, which supports planning for EV charging infrastructure on the nation's interstate highways.

- o Update its assessment of the ability and plans of the private sector to build out the EV charging infrastructure consistent with the pace of EV deployment needed for vehicle electrification anticipated for deep decarbonization, and the need for vehicles on interstate highways and in public locations or high-density workplaces, and to identify gaps in funding and financial incentives as needed.
- DOE, in coordination with FHWA, to provide funding for additional EV infrastructure that would: cover gaps in interstate charging to support long-distance travel and make investments for EV charging for low-income businesses and residential areas.

  Cost: $5 billion.

- The National Institute of Standards and Technology (NIST) to develop communications and technology interoperability standards for all EV Level 2 and DC fast-charging infrastructure.

Broadband

The operational performance and affordability of the low-carbon electricity system will depend on both low-carbon resources as well as flexible demand, with the latter particularly important in an electric system dominated by intermittent generating resources (like solar and wind). Flexible demand, in turn, will depend on the ability of households, businesses, and others to communicate with wholesale and local power markets in real time.

Vast geographic segments of the United States, notably in rural areas and in low-income urban areas, lack access to broadband (Anderson and Kumar, 2019; Perrin, 2019). According to the Federal Communication Commission's (FCC's) most recent report, over 21 million Americans did not have access to high-speed broadband as of the end of 2017 (FCC, 2019), and economic barriers prevent private broadband companies from reaching these communities and inhibit states from providing financial incentives to overcome these barriers. This situation poses countless challenges for millions of households. From the point of view of decarbonizing the nation's energy system, individual electricity customers without broadband cannot effectively respond to price and demand management signals to allow them to play a part in flexible demand strategies. The deployment of advanced meters (addressed earlier) must be accompanied by deployment of broadband to enable that capability. Further, a Brookings Institution analysis indicates that although the FCC provides subsidies to assist rural areas (e.g., $186 million in 2018; see Conexon, 2018), it would take in the range of $14 billion to $28 billion to provide universal broadband access (Levin, 2019). Investment tax credits (or grants to publicly owned utilities for 10 percent of that cost would help the private sector and others to accelerate such deployment.

The committee recommends that Congress enact:

- Statutory changes to enable rural electric cooperatives to invest in broadband technology and projects and to provide communications services to their customer base, with appropriations that would provide for grants and/or loans to public power entities equal to 10 percent of investment costs.

  Cost: $0.5 billion.

- Investment tax incentives (at 10 percent of investment) for private companies to make broadband investments in low-income and rural communications.

  Cost: $1.5 billion.

CO_2 Pipeline Infrastructure

Consistent with the recommendations in Chapter 2 regarding the potential need for on the order of 50–75 MMT CO_2 capture and storage per year by 2030 (predominately at industrial facilities) and as much as 250 MMT CO_2 by 2035, the nation needs to plan and construct a new interstate CO_2 transportation system to move quantities of CO_2 from sources to long-term storage locations. Although there are currently 50 CO_2 pipelines (totaling 4,500 miles) already in existence, they are used primarily to move CO_2 for injection in oil-producing fields to enhance recovery of oil and are insufficient for carbon capture, utilization, and sequestration (CCUS) at this scale (Wallace et al., 2015). The Princeton Net Zero America study (Larson et al., 2020) has modeled CO_2 pipelines required for lowest-cost net-zero energy systems in the United States in a variety of scenarios, the least-constrained and lowest-cost of which would require an additional 16,000 km (or around 10,000 miles) of pipelines before 2030 to facilitate installation of CCUS (Larson et al., 2020).

A recent study by researchers at the Great Plains Institute and the University of Wyoming concluded that it will be more economical to build out that CO_2 delivery infrastructure if it is done in a coordinated fashion:

> A regional network will require coordination between states, possibly coordination between multiple pipeline owners and operators, and long-term planning of likely capture and storage locations to determine routes and expected capacity requirements. A transport network built only with near-term projects in mind will require greater land use and induce higher costs on a

> per ton basis than a regional network planned with a longer time horizon. . . . Long-term, coordinated planning on regional CO_2 transport corridors will result in optimized, regional scale infrastructure that minimizes costs, land use, and construction requirements while maximizing decarbonization across industrial and power sectors throughout the United States. . . . To avoid the business-as-usual and expensive outcomes in which CO_2 transport infrastructure is built out in a piecemeal fashion, . . . planning and coordination must occur in the near term to begin building regional-scale transport networks for economy-wide deployment of carbon capture and storage. (Abramson et al., 2020)

Planning for such a CO_2 transportation network should take place in the next 5 to 10 years, and include public participation and expert input. Such planning should take into account the current and likely future location of large point sources of CO_2 (e.g., above 0.5 Mt CO_2/year) and CO_2 sequestration basins, and seek to enable 95 percent of all current and future likely large point sources of CO_2 to fall within a reasonable distance (e.g., 100 miles) of the trunk-line system. The plan should focus on using, to the extent possible, existing rights-of-way to site CO_2 trunk lines. One recent study matched potential sources and subsurface storage sites for carbon capture and sequestration (CCS) in California (EFI, 2020).

Other elements of planning for CO_2 storage infrastructure involve characterization of reservoirs for safe and permanent storage of CO_2. DOE, in conjunction with the U.S. Geological Survey (USGS) and the Department of the Interior (DOI), should begin to characterize all major basins for CO_2 sequestration in order to identify with high-confidence sites suitable for at least 1 Gt CO_2/year of injection with permanent containment. This effort should be conducted via a highly coordinated public-private partnership that supports exploration and appraisal, field development, extensive stakeholder engagement, plugging and abandonment of legacy wells, and environmental permitting.

Additionally, the regulatory infrastructure to review and approve facilities in this interstate system will need to be established during the next 5 to 10 years. Enhanced technical and legal regulatory capabilities will also be needed (e.g., at EPA, or FERC, or DOT) to review and permit CO_2 injection sites. Congress should establish a National Commission to identify and present recommendations for legislation with regard to legal, policy, and financial considerations related to insurance, public and/or private ownership structure, financing risks, liability issues, regulation, enforcement, and other responsibilities in a CO_2 transportation and sequestration industry.

The committee recommends that Congress:

- Establish a temporary National Commission to identify and present recommendations for legislation related to roles and responsibilities of federal and state agencies and the private sector in a CO_2 transportation and storage industry.

    Cost: $20 million.

- Assign responsibility to DOT, in consultation with DOE, DOI, and EPA, to conduct a planning process for the layout, location, siting principles, and timing of a national CO_2 transportation infrastructure (or "trunk-line" system) to connect sources of CO_2 with locations for permanent sequestration and/or use of CO_2.
  - This planning process must include public participation of communities located near any of the potential routes of CO_2 trunk-line systems, including locations where CCUS projects are likely to be located and locations where CO_2 sequestration would likely occur. By mid-decade (2025–2026), DOT and DOE, in consultation with the other federal agencies, will conduct and publish the results of an assessment to determine the timing of when such a CO_2 trunk-line system would be needed to achieve a net-zero economy by 2050. This report should contain a set of candidate trunk lines, routes, and a timeline for commencement and completion of pipeline segments consistent with the goal of a net-zero economy by 2050. The report should also consider and issue recommendations on what federal financing support, if any, is needed for such a system to be financed, built, and operated, including consideration of what role, if any, the Green Bank should play in supporting such financing.

  Cost: $50 million for planning.

- Appropriate block grants to support community and stakeholder engagement in the planning of the national CO_2 transportation infrastructure above, including staff time for nongovernmental and community organizations to participate.

  Cost: $50 million.

- Direct and fund DOE, USGS, and DOI to characterize with high confidence all major basins for CO_2 sequestration and, by 2030, identify sites suitable for injection of approximately 250 million metric tons of CO_2 per year.

  Cost: $5 billion.

- Establish and fund federal research, development, and demonstration (RD&D) programs to expand technological options for carbon storage and use including the ability of building materials, products, and infrastructure to sequester carbon through bio-materials, carbon fuels, and encapsulation.
- Extend 45Q tax credit for CCUS for projects that begin substantial construction prior to 2030 and make tax credit fully refundable for projects that commence construction prior to December 31, 2022. Set the 45Q subsidy rate for use equal to $35/t$CO_2$ less whatever explicit carbon price is established and the

subsidy rate for permanent sequestration to be equal to $70/tCO_2$ less whatever explicit carbon price is established.

Cost: $2 billion.

Invest in Educational Programs for a Clean Energy Workforce

To navigate the transition to a carbon-neutral economy, the United States needs substantial new investments in education and workforce development. The educational gap across a wide range of clean energy fields (engineering, sciences, architecture and design, construction and facility management, social sciences, public policy and administration, and business and entrepreneurship) is as stark as that which inspired the National Defense Education Act of 1958 after the Soviet Union launched Sputnik, inspiring the International Space Race.

Training the next generation of business, policy, and civil society leaders not only to successfully navigate the complexities of the transition but also to ensure that the United States regains the global lead in energy innovation will require significant new investments. To meet this need, Congress should establish a 10-year GI Bill-type of program to fund vocational, undergraduate, or master's degrees related to clean energy, energy efficiency, building electrification, sustainable design, or low-carbon technology. The Post-9/11 GI Bill has supported approximately 228,000 beneficiaries per year at a cost of approximately $9 billion per year (CBO, 2019). Given the grave threat to the nation posed by climate change and the opportunities presented by a clean energy transition, a program at approximately half the size would position the nation to produce the workforce it needs to confront the threat and take advantage of the opportunity. Such a program would ensure that the U.S. workforce transitions along the physical infrastructure of energy, transportation, and economic systems. It would not only increase the skilled workforce for clean energy, which will require new skills and expertise, and prepare the energy workforce to effectively accommodate transformative technological change in machine learning, big data, automation, and artificial intelligence, but also ensure that the United States remains competitive in rapidly changing global energy markets and trade regimes. To collect the necessary data to understand clean energy workforce needs and gaps, and also to identify and implement ways to address them, the Energy Jobs Strategy Council should be reestablished.

The new GI program for worker training should provide effective and equitable access to good jobs, training (including job placement and/or a pipeline to those jobs) and advancement, particularly for those historically underrepresented or adversely impacted or dislocated by technological change such as energy, transportation, and trade-impacted communities. New educational programs can train an inclusive workforce for high tech, advanced manufacturing, as well as clean energy infrastructure build-out. This investment should also integrate with community services to maximize retention and advancement of workers, particularly disadvantaged or previously underrepresented workers, in clean economy careers. These investments will contribute to more equitable educational attainment in science, technology, engineering, and medicine (STEM) fields (Bound and Turner, 2002), which remains a critical shortcoming of U.S. higher education and an important reason why the benefits of science and technology disproportionately do not flow to low-income communities and communities of color.

Specific attention should be paid to training and providing access to manufacturing occupations to build the skilled workforce to produce the equipment needed for achieving a carbon neutral economy. Manufacturing jobs can provide a pathway to the middle class for workers and families, furthering support for the social contract for decarbonization. Pipelines can be started in high school and on to vocational schools that could have nationally accredited qualifications, making higher paid careers more accessible to lower-income Americans. Ongoing technical and on-the-job training can help workers gain skills, experience, and recognized credentials to advance in their careers. Mobile training labs can be used to bring training to Indigenous peoples and others located in isolated areas.

To meet the needs of these trainees and workers, Congress should support the creation of innovative new degree programs in community colleges and colleges and universities focused uniquely on the knowledge and skills necessary for a low-carbon economic and energy transformation. Too few degree programs, even in energy and environmental studies, provide rigorous training in transition management, and this gap is doubly significant in more traditional programs in engineering, business, policy and administration, which need to be upgraded to ensure graduates are positioned to add new knowledge and skills to their employers. Congress should fund grants to universities at a cost of $100 million per year to create or strengthen undergraduate and master's degree programs in climate- and energy-transition-related studies, whether in engineering, design and architecture, social sciences, natural sciences, or public policy.

Last, Congress should also make significant new investments at the master's, doctoral, and postdoctoral levels to support clean energy innovation. Expanding the number of academic institutions awarding doctorates related to clean energy (engineering, sciences, architecture and design, social sciences, public policy and administration) should be a priority. Congress should also provide grants of $50 million per year to create interdisciplinary doctoral and postdoctoral training programs, similar to those funded by the National Institutes of Health (NIH), which place an emphasis on training students to pursue interdisciplinary, use-inspired research in collaboration with external stakeholders that can guide research and put it to use in improving practical actions to support decarbonization and energy justice.

Overall (not exclusive to clean energy), university-based research, skill formation, and knowledge generation is highly concentrated. Just 115 U.S. universities perform three-quarters of all academic R&D and also award three-quarters of U.S. science and engineering doctoral degrees (NSF, 2020). Congress should provide $375 million per year to support government-funded doctoral and postdoctoral fellowships in science and engineering, policy, and social sciences, for students researching and innovating in low-carbon technologies, sustainable design, and energy transitions, with at least 75 fellowships per state to ensure regional equity and build skills and knowledge throughout the country. Allocation of scholarships must ensure that students of all backgrounds can pursue their passions. These scholarships should include appropriate training in skills in interdisciplinary research and communication, as well as collaboration with industry, government, and civil society stakeholders, in order to ensure that researchers are prepared to work effectively in teams on use-inspired research that contributes meaningfully to the needs of society and the economy.

In the past, the United States has had a comparative advantage through its ability to recruit and retain talent in its high-tech industries from around the world. Studies have shown that the recruitment of foreign graduate students to the United States has had a significant and positive impact on innovation as measured by both future patent applications and future patents awarded to university and non-university institutions (Chellaraj et al., 2008; Hunt and Gautheir-Loiselle, 2010). The United States must redouble efforts to attract talent in low-carbon energy. Visa restrictions for international students who want to study climate change and clean energy at the undergraduate and graduate levels should be eased or eliminated, where appropriate.

The committee recommends that:

- Congress should establish a 10-year GI Bill-type program for anyone who wants a vocational, undergraduate, or master's degree related to clean energy, energy efficiency, building electrification, sustainable design, or low-carbon technology. These programs should include a cost-of-living stipend. Such a program would ensure that the U.S. workforce transitions along the physical infrastructure of our energy, transportation, and economic systems.

    Cost: $5 billion/year for 10 years.

- Congress should support the creation of innovative new degree programs in community colleges and colleges and universities focused uniquely on the knowledge and skills necessary for a low-carbon economic and energy transformation.

    Cost: $100 million/year.

- Congress should also provide funds to create interdisciplinary doctoral and postdoctoral training programs, similar to those funded by NIH, which place an emphasis on training students to pursue interdisciplinary, use-inspired research in collaboration with external stakeholders that can guide research and put it to use in improving practical actions to support decarbonization and energy justice.

  Cost: $50 million/year.

- Congress should provide support for doctoral and postdoctoral fellowships in science and engineering, policy, and social sciences, for students researching and innovating in low-carbon technologies, sustainable design, and energy transitions, with at least 25 fellowships per state to ensure regional equity and build skills and knowledge throughout the United States.

  Cost: $375 million/year.

- The Department of Homeland Security should eliminate or ease visa restrictions for international students who want to study climate change and clean energy at the undergraduate and graduate levels, where appropriate.
- Congress should pass the Promoting American Energy Jobs Act of 2019 to reestablish the Energy Jobs Strategy Council under DOE, require energy and employment data collection and analysis, and provide a public report on energy and employment in the United States.

  Cost: $7 million over the 2020–2025 period (CBO, 2020).

Invest in a Revitalized Manufacturing Sector

The United States cannot gain global market share in clean energy industries if it does not produce clean energy technologies. Yet, the global market for clean energy is already immense and growing, and U.S. firms and workers are being left behind. The International Energy Agency (IEA) estimates that the global market for clean energy technologies will be $2 trillion during the 5-year period between 2020 and 2025 (IEA, 2019).

The United States can revitalize domestic manufacturing through smart and targeted industrial policies, including establishment of predictable and broad-based market formation policies (such as carbon taxes, performance standards, and tax credits that create demand for low-carbon goods and services), improving access to finance, creation of performance-based manufacturing incentives (including efficiency standards), and export promotion. Inconsistent and volatile policies will fail to revitalize the manufacturing sector because manufacturing firms cannot count on them. Firms must literally be able to capitalize on policies that create markets for low-carbon goods and services, and they cannot do that if policies are unstable and volatile. Firms must be able to demonstrate to financiers that a clear return on investments in production and workers is possible because a market for low-carbon products and services will certainly exist in the United States.

The U.S. government should provide manufacturing incentives to firms that are matched with corresponding performance requirements. Subsidies provided directly to manufacturers must be tied to the meeting of performance metrics, such as the achievement of production and export targets or meeting labor, efficiency, and environmental standards. Manufacturers should also be required to develop strategies for assuring the availability and resilience of their supply chain.

The main policy tools available include loans, loan guarantees, tax credits, export-promotion, and grants to manufacturers, some of which could be administered through the Green Bank, if established. The least costly to the taxpayer is the loan guarantee, which was used successfully during the American Reinvestment and Recovery Act. This program should be reformed to support new and additional advanced technologies, to finance more small and medium-size enterprises, and to encourage more risk-taking on the part of DOE. In export promotion, the U.S. Export-Import Bank needs to phase out support for fossil fuels and make support for clean energy technologies a top priority. U.S. export credit authorizations for renewables have fallen from $200 million in 2014 to just $19 million in 2019 (Ex-Im Bank, 2019). The committee recognizes that each of these policy approaches has limitations. Large corporations can already secure advantaged loan rates, thus loans may be best for small and medium-size manufacturers. Tax credits face limitations, because many companies have already taken the maximum amount of tax credits they can afford to take. Thus, the committee believes that all of these policy tools are necessary.

The committee recommends that Congress:

- Establish predictable and broad-based market-formation policies that create demand for low-carbon goods and services, improve access to finance, create performance-based manufacturing incentives, and promote exports.
- Provide manufacturing incentives through loans, loan guarantees, tax credits, grants, and other policy tools to firms that are matched with corresponding performance and wage requirements. Subsidies provided directly to manufacturers must be tied to the meeting of performance metrics, such as production of products with lower embodied carbon or adoption of low-carbon technologies and approaches. Specific items could include the following:
    - Expand the scope of the energy audits in the DOE Better Plants program and expanded technical assistance to focus on energy use and GHG emissions reductions at the 1,500 largest carbon-emitting manufacturing plants.

- ○ Support the hiring of industrial plant energy managers by having DOE provide manufacturers with matching funds for 3 years to hire new plant energy managers.
- ○ Enable the development of agile and resilient domestic supply chains through DOE research, technical assistance, and grants to assist manufacturing facilities address supply chain disruptions resulting from COVID-19 and future crises.

Cost: Initial appropriation of $1 billion/year phasing down over 10 years as performance targets are reached.

- Provide loans and loan guarantees to manufacturers to produce low-carbon products, ideally through a Green Bank.
- Require the U.S. Export-Import Bank to phase out support for fossil fuels and make support for clean energy technologies a top priority with a minimum of $500 million/year.
- Create a new Assistant Secretary for Carbon Smart Manufacturing and Industry within DOE.

Invest in Research, Development, and Demonstration for Technology Innovation and Deployment and Research on Social and Economic Impacts

The United States needs to dramatically strengthen its knowledge base on clean-energy technologies as well as on the social dimensions of transitions to a net-zero carbon economy. Such investments require increased federal support.

American public investments in clean energy technology RD&D have gradually risen since 2011 but U.S. leadership in clean energy RD&D is now being challenged by China and Europe. The United States led the world in public investments in clean energy RD&D from the 1970s until the late 2010s when China's public investments began to rival or even exceed U.S. investments. China will likely double government RD&D spending on clean energy between 2015 and 2020 from $4 billion to $8 billion (Myslikova and Gallagher, 2020). This achievement will put China's officially reported RD&D spending on clean energy ahead of that of the United States. U.S. investments in clean energy RD&D increased by 42 percent between 2015–2020 from $4.8 billion to $6.8 billion (including basic energy sciences) owing to sustained support from congressional appropriations, despite the Trump administration's proposed drastic cuts of more than 60 percent to clean energy RD&D every year in its budget request to Congress (Myslikova and Gallagher, 2020). European clean energy RD&D investments as of 2018 were approximately $6.3 billion.

To restore U.S. leadership in clean energy technology RD&D, the committee recommends that Congress triple the DOE's funding of low- or zero-carbon RD&D over the next 10 years, in part by eliminating investments in fossil-fuel RD&D. A tripling of energy innovation investments was recommended by the American Energy Innovation Council in 2020 (AEIC, 2020), Sivaram et al. (2020), and by the President's Council of Advisors on Science and Technology in 2010 (PCAST, 2010). Other recommendation to greatly increase DOE's funding of clean energy technologies include the call by Nobel Prize winners to the Obama administration (Burton, 2009) and the testimony of eventual DOE Secretary Ernest Moniz during his confirmation hearing (S. Hrg. 113-17, 113th Cong., 2013).These investments should focus on the five critical actions discussed in Chapter 2 as well as the technologies that need to be better understood for possible deployment in the 2030s, including clean firm electricity resources, buildings and industrial efficiency, electricity storage, CCS, hydrogen and other low or net-zero carbon energy carriers, high-yield bioenergy crops, low-emissions industrial process technologies, and negative emission technologies (NETs). By eliminating investments in non-CCS fossil-fuel RD&D, the net increase to the energy RD&D budget will be partially offset. DOE should also fund energy innovation policy evaluation studies to better understand the extent to which policies implemented (both RD&D investment and market-formation policies) are working. Relatedly, DOE and/or the National Science Foundation (NSF) should support studies on the socioeconomic impacts of low-carbon transitions.

As funding ramps up, Congress should target under-resourced sectors and gaps in the U.S. innovation system for the largest increases. The end-use sectors are particularly under-represented in the current RD&D portfolio (Sivaram et al., 2020; IEA, 2020; Shah and Krishnaswami, 2019; Breakthrough Energy, 2019). Sivaram et al. (2020) find that less than a quarter of DOE's portfolio targets innovations in the transportation, buildings, and industrial sectors. The IEA (2020) recommends that the world's major economies provide more funding for end-use innovations in sectors such as heavy industry and long-distance transportation that have no or few commercially available low-carbon options.

It is important to note that there is critical gap in government funding between basic research and commercialization. For example, while the Advanced Research Projects Agency-Energy (ARPA-E) has been successful in the development of innovative technologies, the National Academies review of the agency noted that none of these innovations has resulted in new commercial technologies (NASEM, 2017a). Other reviews of ARPA-E have noted this same gap (Goldstein et al., 2020), and national laboratories face similar difficulties in moving innovations to commercial products (Stepp et al., 2013; Anadon et al., 2016; Chan et al., 2017). One method being explored is to scale up funding for entrepreneurial research fellows, which is showing promise in the current lab-embedded entrepreneurship program (LEEP) configuration, such as Cyclotron

Road at the Lawrence Berkeley National Laboratory or Chain Reaction at Argonne National Lab. Similarly, the Small Business Voucher Pilot Program, launched in 2015 to increase small business access to lab capabilities, was successful at helping small businesses advance their technologies and achieve commercial sales (Jordan and Link, 2018). Programs such as these that increase private-sector access to federal research facilities, as well as incentives that encourage research staff to collaborate with industry, should be expanded.

Successfully shepherding new technologies from concept to commercialization requires support at all stages, but the demonstration stage is particularly underfunded (C2ES, 2019; Nemet et al., 2018; Hart, 2018). The IEA defines technology demonstration as the "operation of a prototype . . . at or near commercial scale with the purpose of providing technical, economic and environmental information" (IEA, 2011). The fundamental role of demonstration is to instill confidence in technology developers, users, investors, and the public that a technology will perform as intended. However, the first several large demonstrations of an emerging technology generally entail a level of technical and financial risk beyond what private industry can support, leading to a "commercialization valley of death" (Nemet et al., 2018).

The federal government virtually stopped funding demonstrations after the American Recovery and Reinvestment Act of 2009 expired. Today, the only federal funding for demonstration projects is under a new program for advanced nuclear reactors, which was approved by Congress in FY 2020 (U.S. Congress, Senate, 2019). The Title XVII Loan Guarantee Program provides some support for first-of-a-kind commercial projects that could include demonstrations. But loan guarantees on their own may not be sufficient to induce the private sector to invest in novel technology demonstrations. Green banks—which are generally expected to retain their initial capital and therefore require a return on their investments—are similarly ill-suited for large demonstrations (Rozansky and Hart, 2020).

Congress has repeatedly affirmed its support for later-stage R&D and demonstration activities (H. Rep. 116-83, 116th Cong., 2019; U.S. Congress, Senate, 2019), but demonstrations remain a critically underfunded portion of the federal energy innovation portfolio (Rozansky and Hart, 2020; Krishnaswami and Higdon, 2020; Sivaram et al., 2020). The American Energy Innovation Act introduced in the Senate in February 2020 would require DOE to conduct 17 demonstration projects across four technology areas: energy storage, carbon capture, enhanced geothermal systems, and advanced nuclear (U.S. Congress, Senate, 2020). But demonstrations across a broader range of technologies will be necessary to address the full range of innovation needs. Within the RD&D portfolio, Congress should increase funding for demonstration projects.

Meanwhile, the softer costs (e.g., permitting, interconnection) of clean energy remain higher in the United States than in other countries, indicating that there is still room for final cost reductions for clean energy technologies. Therefore, DOE should fund studies aimed at reducing the soft costs of zero-carbon technology, including through policy.

The private sector is a major contributor to U.S. energy RD&D, but owing to the lack of reporting, it is unclear how much firms are investing in clean energy RD&D and whether public investments duplicate private investments. Thus, the committee recommends that all firms receiving funds from the government be required to report on their aggregate investments in RD&D annually, by type of investment (basic energy sciences, applied RD&D) and category (e.g., solar, wind, smart grid, fission, fusion, negative emission, efficiency). Additionally, the committee recommends that such RD&D expenditures be disclosed in corporate filings to the Securities and Exchange Commission.

Certain public-private partnerships (PPPs) have been successful for DOE in the past, and those should be studied with a view to enhancing PPPs in clean energy. Relatedly, low-carbon advanced manufacturing capabilities should be bolstered through PPPs for RD&D on advanced manufacturing in clean energy, the establishment of government-sponsored platforms for demonstration of improved manufacturing techniques, and establishment of regional innovation and manufacturing hubs for low-carbon energy around the country.

DOE should establish regional innovation hubs where they do not yet exist to focus involvement of the private sector and state, private, and rural colleges and universities and national laboratories. These regional innovation hubs should be focused on deep energy efficiency activities (e.g., ones that could reduce a building's energy consumption by 50 percent or more) and the development and exploitation of clean energy resources where there is a comparative advantage for that region (e.g., solar in the Southwest, offshore wind in the Northeast, onshore wind in the upper Midwest).

The committee recommends that Congress should:

- Triple DOE's government investments in low- or zero-carbon RD&D over the next 10 years, in part by eliminating investments in fossil-fuel RD&D. These investments should include renewables, efficiency, storage, transmission and distribution, CCUS, advanced nuclear, and NETs and increase the agency's funding of large-scale demonstration projects. By eliminating investments in non-CCS fossil-fuel RD&D, the net increase to the energy RD&D budget will be partially offset.

    Cost: Increase from $6.8 billion per year in 2020 to $20 billion per year by 2030, but partially offset by eliminating the non-CCUS fossil budget, which for FY 2020 is $273 million for coal and $15 million for gas and unconventional, which would be $2.8 billion over 10 years.

- Direct DOE to fund energy innovation policy evaluation studies so that the extent to which policies implemented (both RD&D investment and market-formation policies) are working.

    Cost: $25 million/year.

- Direct DOE and NSF to create a joint program to fund studies of the social, economic, ethical, and organizational drivers, dynamics, and outcomes of the transition to a carbon-neutral economy, as well as studies of effective public engagement strategies for strengthening the U.S. social contract for decarbonization. Such studies should improve the understanding of how large-scale energy transitions can be accomplished; the full complexity of the diverse scientific, industry, and societal innovation systems involved; the factors that contribute to accelerating or delaying processes of change; and the rich intersections between changes in energy technologies and social practices and other processes of social and economic change.

    Cost: $25 million/year.

- Direct DOE to establish regional innovation hubs where they do not exist or are critically needed using funds appropriated in tripling DOE's government investments in low- or zero-carbon RD&D.

    Cost: $20 million/year.

- Direct DOE to enhance public-private partnerships for low-carbon energy.

Invest in Efficiency Improvements for Low-Income Households Through Program Redesign and Expanded Funding

High energy burdens and lack of capacity to invest in infrastructure improvements work to reinforce energy and economic insecurity for many low-income households, small businesses, and communities in the United States. In some cases, total energy costs can be as high as 25 percent or more of monthly income, especially when electricity, natural gas, and gasoline costs are included.

The two principal federal programs to assist in lowering low-income consumers' energy bills are the Weatherization Assistance Program (WAP) and the Low-Income Home Energy Assistance Program (LIHEAP), both of which are administered by the states. DOE has responsibility for WAP, and the Department of Health and Human Services (HHS) manages LIHEAP, with HHS allowing each state to use up to 15 percent of LIHEAP dollars to add to WAP funding (and up to 25 percent with an approved "good cause" waiver). Although low-income customers benefit in the short term from the assistance they receive in paying their energy bills, the federal government should expand on the ability to use federal dollars to leverage long-term efficiency investments to lower bills for years to come. As part of the ARRA economic stimulus funding, Congress expanded WAP funding from $236 million in 2008 to $2 billion in 2010; this provided substantial energy savings in the nation's residential buildings, saved each participating household thousands of dollars in energy bills, provided them with health and safety benefits, and produced thousands of jobs in local communities (Tonn et al., 2015). In addition, the USDA's Rural Energy Savings Program allows consumers to finance energy-saving home improvements with no upfront costs through rural electric co-ops. Loans are paid back over time with savings resulting from the consumers' reduced energy consumption.

Congress should increase the combined dollars that go to LIHEAP and WAP, as various analyses have indicated the success of these programs (Murray and Mills, 2014; Fowlie et al., 2018; Tonn et al., 2018; Terman, 2018), allow the states to request approvals of using a higher percentages of LIHEAP dollars (up to 25 percent across the board, and up to 35 percent with a good cause waiver), and encourage states to coordinate WAP grants to households with other energy-efficiency programs funded by utilities and their customers. Specifically, expanded funding from the WAP program should also fund electrification of buildings' heating and cooling systems, and include financial support for low-income communities (e.g., through local hiring requirements, local supply sourcing, or other approaches, to ensure that local communities benefit from the employment and spending associated with these programs).

The committee recommends that Congress:

- Expand funding of the WAP program to $12 billion over the next the next 10 years (front-loading spending to get the benefits as soon as possible), without reducing funding for LIHEAP, and direct HHS to allow states to use a greater share of LIHEAP dollars for investments in energy efficiency measures and electric heating and cooling systems.

    Cost: $1.2 billion/year for 10 years.

Invest in Electrification of Tribal Lands

Access to electricity is critical for improving standards of living, education, and health (U.N. Development Programme, 2019), but as many as 160,000 Native Americans still lack access to electricity (DOE, 2017). In the Navajo Nation alone, about 15,000 homes have no electricity (DOE, 2018). More than 175 remote Alaska Native villages are not connected to a larger electricity grid and rely on imported diesel fuel for electricity generation, resulting in electricity costs as high as $1.00/kWh—8 times the national average (Schwabe, 2016). A poll conducted by NPR, the Robert Wood Johnson Foundation, and the Harvard T.H. Chan School of Public Health (2019) found that more than a quarter of Native Americans have experienced problems with electricity, internet access, and safe drinking water. About one in four Native Americans lives in poverty, with unemployment rates twice as high as those among non-Native Americans nationally (DOE, 2020c).

DOE (2017) found that "it is a moral imperative that the federal government support tribal leadership and utility authorities to provide basic electricity service for the tens of thousands of Native Americans who currently lack access to electricity and to foster the associated economic development on tribal lands," and recommended that federal agencies support full tribal land electrification. However, electrification of tribal lands faces significant challenges. For example, the low population density in the Navajo Nation means the connection cost is as high as $40,000 per home (NPR, 2019).

Current federal programs to support tribal electrification include the DOE Office of Indian Energy (DOE-IE), which provides financial and technical assistance, and the DOI Bureau of Indian Affairs (DOI-BIA), which provides support for strategic development and project planning. Additionally, USDA's Rural Utilities Service offers low-cost loans to rural utilities and tribal authorities to expand grid access. The Energy Policy Act of 2005 authorized the DOE Tribal Energy Loan Guarantee Program (TELGP) to provide up to $2 billion in partial loan guarantees to support energy development projects. However, Congress did not appropriate funding for the credit subsidy until fiscal year 2017. As of February 2020, DOE had not issued a single tribal energy loan guarantee (DOE, 2020c).

Congress should increase funding for tribal electrification programs at DOE, DOI, and USDA to enable full electrification by 2030, while respecting the sovereignty of tribal and Alaska Native communities. DOE-IE and DOI-BIA should provide technical assistance in long-term planning, project development, legal and regulatory assistance, and siting and permitting assistance for projects. Additionally, Congress should increase direct financial assistance for the buildout of electricity infrastructure through DOE-IE grant programs.

The committee recommends that Congress:

- Provide $20 million per year over the next 5 years for needs assessment, strategic development, and planning through DOE-IE grants and the USDA Rural Utilities Service (USDA-RUS) High Energy Cost Grant Program.

    Cost: $20 million/year for next 5 years.

- Expand funding of the DOE-IE financial assistance program to $200 million per year over the next 10 years, and amend the Rural Electrification Act to allow USDA-RUS to lend at 0 percent interest through the Substantially Underserved Trust Areas program.

Cost: $200 million/year for next 10 years.

STRENGTHENING THE U.S. CAPACITY TO EFFECTIVELY AND EQUITABLY TRANSITION TO A CLEAN ENERGY FUTURE

A just, equitable, effective, and rapid transition to a carbon neutral economy in the United States will require significantly improved coordination of planning and action within and across various levels of decision making, including local, state, and federal governments and countless other stakeholders in industry and civil society. This extensive coordination is essential to properly design and implement accelerated technological changes toward carbon neutrality and also to ensure that the resulting economic and societal transformation advances the broad goals identified in Chapter 3 and meets the benchmarks for equity and inclusion established and monitored by the White House Office of Equitable Energy Transition (recommended earlier in this chapter). This section describes the policies needed to enable institutions to manage and plan the transition.

The committee emphasizes that strengthened coordination is especially required to address several key features of the transition to decarbonization. The first is the extensive and complex interactions between the energy system and multiple sets of critical infrastructures, including but not limited to manufacturing, transportation, food, water, communication and information, supply chains, housing, and security. Many of these systems depend on public and private investments and governance structures affected by markets and multiple layers of government. The second is the tight coupling of energy systems operations, performance, supply chains, and regulation across local, state, regional, national, and global scales, much of which will need to be adjusted and reoptimized during the transition process. The third is the need for careful attention to ensuring that the broad goals identified in Chapter 3 are met throughout the transition, including rebuilding a strong U.S. economy, ensuring a broad distribution of economic success across the diverse U.S. regions, actively promoting equity and justice for diverse communities, and ensuring that harms created by the transition itself are appropriately anticipated, assessed, and mitigated.

To help facilitate an energy transition that anticipates and addresses these challenges, the committee recommends that the federal government support significantly enhanced planning and coordination efforts across the various levels of government.

At least three impediments stand in the way of accelerated action to advance a just and equitable transition to net zero: a shortage of human and financial resources for planning and coordination; a lack of existing coordination mechanisms and processes at appropriate scales; and a mismatch between existing knowledge resources about low-carbon energy technologies and transitions and the needs of diverse decision makers and other stakeholders. In particular, many local actors, governments, and nongovernmental organizations (NGOs) do not have the capacity or ability to access federal funds, determine what to apply for, or know how to implement the funds for impact. Funding for technical assistance should be provided for local planners, public and private, who know and understand the community and are skilled at accessing and implementing funds for impactful uses.

Congress should act to address these gaps by establishing and funding a multiscale planning infrastructure at federal, state, regional, and local levels with both the capacity to plan and coordinate an accelerated transition and to secure the knowledge resources necessary for that work.

The committee recommends:

- **Federal:** The bulk of the effort at the federal level is described earlier under the sections describing the National Transition Task Force and Office on Equitable Energy Transitions. Efforts from those entities should be focused on instituting better information, analysis, and coordination on issues related to equitable energy transitions. In addition:
    - Congress should direct a portion of federal energy research, development, demonstration, and deployment spending at DOE to provide usable and use-inspired social-science and techno-economic knowledge for decision makers at all levels to support their efforts to plan and implement accelerated actions toward a carbon-neutral U.S. economy. As part of this effort, Congress should provide additional annual funding to the national laboratories to establish a coordinated, multilaboratory capability to provide energy modeling, data, and analytic and technical support to cities, states, and regions to complete a just, equitable, effective, and rapid transition to net zero. This funding should commence at the level of $200 million per year, rising to $500 million/year by 2025, and $1 billion/year by 2030.

    Cost: $4.5 billion over 10 years.

- **Regional:** Congress should create a regional planning and coordination infrastructure to support regional efforts to accelerate the equitable energy transition.

Major U.S. energy, transportation, and economic systems vary significantly across regions and are often organized and governed in regional, multistate arrangements according to regional priorities. Considerable work involved in coordinating, planning for, and managing the transition to a carbon neutral U.S. economy will therefore be necessary at the regional scale. Historically, regional authorities have played important roles in rapid energy system transformation in the past, including during the Depression-era New Deal, and offer the right scale and coordinating function to address the needs of deep decarbonization (Wiseman, 2011). Regional planning offers a mechanism for strengthening the capacity of localities and communities to successfully navigate transitions, to build relationships and work collaboratively with state and federal actors to implement strategic planning, and to integrate energy system planning and economic development (Healey, 1998; Morrison, 2014).

Congress should therefore establish 10 regional transition coordination offices under the auspices of the U.S. Department of Commerce, with advisory assistance from the White House Office of Equitable Energy Transitions, with the mandate to

- Coordinate federal agency actions at the regional scale through the deployment of federal agency staff to regional offices with specific attention and funding for local technical assistance.
- Host a coordinating council of regional governors and mayors that meets annually to establish high-level policy goals for the transition.
- Establish mechanisms for ensuring the effective participation of low-income communities, communities of color, and other disadvantaged communities in regional dialogue and decision making about the transition to a carbon-neutral economy.
- Provide information annually to the White House Office of Equitable Energy Transitions detailing regional progress toward decarbonization goals and benchmarks for equity.

Cost: $5 million/year for each regional office to provide funding for coordinating and hosting meetings, reporting, and information dissemination.

Congress should also:

- Provide $25 million per year for a multi-university collaborative research center in each region to provide the data, models, and social science needed by regional transition coordination offices and local and state organizations to successfully navigate the complexities of regional transitions to net zero.

These centers should be funded and administered through a competitive grant-making process coordinated by the National Science Foundation, with clear guidance regarding required collaboration with local, state, and regional stakeholders to set research agendas, design research, and disseminate research findings.

Cost: $25 million/year.

- **State:** Congress should encourage each state to accelerate and coordinate the decarbonization of its economy. To accomplish this, Congress should direct DOE to:
    - Provide up to $1 million per year in matching funds to establish in each state an office of equitable energy transition in the governor's office or other cross-agency senior administrative position. This office will coordinate state efforts to accelerate the transition of the state's economy to carbon neutrality, host statewide stakeholder and community councils to coordinate decarbonization efforts, and coordinate state participation in regional transition coordinating councils. The office will also provide information to the Office of Equitable Energy Transitions on state progress toward carbon neutrality, the societal and economic criteria identified in Chapter 3, and the benchmarks established by the Office of Equitable Energy Transitions. The office will also establish mechanisms for ensuring the effective participation of low-income communities, communities of color, and other disadvantaged communities in state dialogue and decision making about the transition to a carbon-neutral economy consistent with standards set by the Office of Equitable Energy Transitions.

    Cost: Up to $50 million/year.

- **Local:** The capacity of cities and counties to pursue planning has been severely undermined by the erosion of state and local budgets during and after the recession of 2008–2010. COVID-19 has compounded these challenges, further reducing city and county finances and staffing. These impacts pose severe challenges to the ability of municipalities and communities to pursue the scale and depth of planning necessary to ensure successful decarbonization by 2050.

    Congress should therefore provide incentive-based financial support and local technical assistance to municipal and county governments to create and strengthen local processes for planning decarbonization. These planning processes should (1) ensure coordinated planning at the local level across sectors and communities; (2) remove local planning barriers to accelerating actions to promote decarbonization and meet societal and economic criteria; (3) provide annual progress reporting; and (4) enable proactive identification of vulnerable communities, assess the challenges they face, and ensure their effective participation in transition planning. To create these incentives, Congress should:

- Fund $1 billion per year in community block grants to support local decarbonization planning through a federal grant-making program. The grant-making program would be funded through DOE, while the grants would be administered and synthesized through the regional transition coordination office with local technical assistance for the region where the community is located.

Cost: $1 billion/year.

- Include provisions so that the block grants include appropriate processes and allocation of resources to ensure inclusive, effective engagement and participation of low-income communities, communities of color, and other disadvantaged communities in planning processes.

REFERENCES

Abramson, E., D. McFarlane, and J. Brown. 2020. *Transport Infrastructure for Carbon Capture and Storage: Whitepaper on Regional Infrastructure for Midcentury Decarbonization*. Minneapolis, MN: Great Plains Institute and University of Wyoming.

AEIC (American Energy Innovation Council). 2020. *Energy Innovation: Supporting the Full Innovation Lifecycle*. Washington, DC. February.

Aggarwal, S., S. Corneli, E. Gimon, R. Gramlich, M. Hogan, R. Orvis, and B. Pierpont. 2019. *Wholesale Electricity Market Design for Rapid Decarbonization*. San Francisco, CA: Energy Innovation.

Ahlquist, J.S. 2017. Labor unions, political representation, and economic inequality. *Annual Review of Political Science* 20: 409–432.

Aldy, J.E. 2011. "Promoting Clean Energy in the American Power Sector." The Hamilton Project Discussion Paper 2011-04.

Aldy, J.E. 2017. Designing and updating a U.S. carbon tax in an uncertain world. *Harvard Environmental Law Review Forum* 41: 28–40.

Aldy, J.E. 2020. Carbon tax review and updating: Institutionalizing an act-learn-act approach to U.S. climate policy. *Review of Environmental Economics and Policy* 14(1): 76-94.

Aldy, J.E., and W.A. Pizer. 2015. The competitiveness impacts of climate change mitigation policies. *Journal of the Association of Environmental and Resource Economists* 2(4): 565–595.

Aldy, J.E., M. Hafstead, G.E. Metcalf, B.C. Murray, W.A. Pizer, C. Reichert, and C. Williams III. 2017. Resolving the inherent uncertainty of carbon taxes. *Harvard Environmental Law Review Forum* 41: 1–13.

Alliance Commission on National Energy Efficiency Policy. 2013. *The History of Energy Efficiency*. Washington DC: Alliance to Save Energy.

Anadon, L., C. Chan, A. Bin-Nun, and V. Narayanamurti. 2016. The pressing energy innovation challenge of the U.S. National Laboratories. *Nature Energy* 1.

Anderson, M., and M. Kumar. 2019. "Digital divide persists even as lower-income Americans make gains in tech adoption." Pew Research Center. https://www.pewresearch.org/fact-tank/2019/05/07/digital-divide-persists-even-as-lower-income-americans-make-gains-in-tech-adoption.

Architecture 2030. 2014. *Roadmap to Zero Emissions*. Amended Version. Santa Fe, NM.

Baker III, J., M. Feldstein, T. Halstead, N.G. Mankiw, H.M. Paulson, G.P. Shultz, T. Stephens, and R. Walton. 2017. *The Conservative Case for Carbon Dividends*. Washington DC: The Climate Leadership Council.

BEA (U.S. Bureau of Economic Analysis). 2018. "Value Added by Industry as a Percentage of Gross Domestic Product, 2018 data." https://www.bea.gov/news/2019/gross-domestic-product-industry-fourth-quarter-and-annual-2018.

Bergquist, P., M. Mildenberger, and L. Stokes, 2020. Combining climate, economic, and social policy builds public support for climate action in the US. *Environmental Research Letters* 15(5).

Bielen, D., D. Burtraw, and K. Palmer. 2017. *The Future of Power Markets in a Low Marginal Cost World*. Washington, DC: Resources for the Future.

Bivens, J. 2015. *A Comprehensive Analysis of the Employment Impacts of the EPA's Proposed Clean Power Plan*. Washington, DC: Economic Policy Institute.

Bivens, J., L. Engdahl, E. Gould, T. Kroeger, C. McNicholas, L. Mishel, Z. Mohkiber, et al. 2017. *How Today's Unions Help Working People*. Washington DC: Economic Policy Institute.

Blonz, J., B. Laird, and K. Palmer. 2018. The benefits of flexible policy design: US energy conservation standards for appliances. *Resources Magazine*. April 6.

Bound, J., and S. Turner. 2002. Going to war and going to college: Did World War II and the G.I. Bill increase educational attainment for returning veterans? *Journal of Labor Economics* 20(4): 784–815.

Breakthrough Energy. 2019. *Advancing the Landscape of Clean Energy Innovation*. https://www.breakthroughenergy.org.

Brooks, S., and N.O. Keohane. 2020. The Political economy of hybrid approaches to a U.S. carbon tax: A perspective from the policy world. *Review of Environmental Economics and Policy* 14(1).

Brückmann, G. and T. Bernauer, 2020. What drives public support for policies to enhance electric vehicle adoption? *Environmental Research Letters* 15.

Burton, T. 2009. "34 Nobel Prize Winners Write President Obama Urging Support for Clean Energy R&D." The Breakthrough Institute. http://thebreakthrough.org.

Burtraw, D., K. Palmer, and D. Kahn. 2010. A symmetric safety valve. *Energy Policy* 38(9): 4921–4932.

Burtraw, D., A. Keyes, and L. Zetterberg. 2018. *Companion Policies under Capped Systems and Implications for Efficiency— The North American Experience and Lessons in the EU Context*. Washington DC: Resources for the Future.

California Climate Investments. 2020. *Annual Report to the Legislature on California Climate Investments Using Cap-and-Trade Auction Proceeds*. Sacramento CA: California Energy Commission.

CA DGS (California Department of General Services). 2018. "Buy Clean California Act." https://www.dgs.ca.gov/PD/Resources/Page-Content/Procurement-Division-Resources-List-Folder/Buy-Clean-California-Act.

California. 2020a. "CALGreen." https://www.dgs.ca.gov/BSC/Resources/Page-Content/Building-Standards-Commission-Resources-List-Folder/CALGreen#@ViewBag.JumpTo.

California. 2020b. "Zero-Emission Vehicle Program | California Air Resources Board." https://ww2.arb.ca.gov/our-work/programs/zero-emission-vehicle-program/about.

CBO (Congressional Budget Office). 2013. *Effects of a Carbon Tax on the Economy and the Environment*. Washington, DC.

CBO. 2019. *The Post-9/11 GI Bill: Beneficiaries, Choices, and Cost*. Washington, DC.

CBO. 2020. "Cost Estimate." https://www.cbo.gov/system/files/2020-02/s2508.pdf.

Chan, G., A.P. Goldstein, A. Bin-Nun, L. Diaz Anadon, V. Narayanamurti. 2017. Six principles for energy innovation. *Nature* 552 (7683).

Chellaraj, G., K.E. Maskus, and A. Mattoo. 2008. The contribution of international graduate students to US innovation. *Review of International Economics* 16(3): 444–462.

Clark, C.E. 2019. *Department of Energy Appliance and Equipment Standards Program*. Washington DC: Congressional Research Service.

Cleary, K., K. Palmer, and K. Rennert. 2019. *Clean Energy Standards: Exploring the options available for policy makers to implement a CES at the state or federal level*. Washington, DC: Resources for the Future.

Clements, A. 2017. "Market Reform to Facilitate Public Policies and a Changing Resource Mix By Allison Clements." *The Future of Centrally-Organized Wholesale Electricity Markets*. Berkeley, CA: Lawrence Berkeley National Laboratory.

Climate Leadership Council. 2020. *The Baker Shultz Carbon Dividends Plan: Bipartisan Climate Roadmap*. Washington, DC.

Climate-Related Market Risk Subcommittee. 2020. *Managing Climate Risk in the U.S. Financial System*. Washington, DC: U.S. Commodity Futures Trading Commission.

Coalition for Green Capital. 2019. *Mobilizing $1 Trillion Towards Climate Action: An Analysis of the National Climate Bank*. Washington DC.

Cohen, J.E. 1995. *How Many People Can the Earth Support?* New York, NY: W.W. Norton and Co.

Collantes, G., and D. Sperling. 2008. The origin of California's zero emission vehicle mandate. *Transportation Research Part A: Policy and Practice* 42: 1302–1313.

Conexon. 2018. "Rural Electric Cooperative Consortium Awarded $186M in FCC's Connect America Fund Phase II Auction, Becoming Single Largest Gigabit Winning Bidder." *PR Newswire*. https://www.prnewswire.com/news-releases/rural-electric-cooperative-consortium-awarded-186m-in-fccs-connect-america-fund-phase-ii-auction-becoming-single-largest-gigabit-winning-bidder-300703739.html.

Corneli, S. 2018. *Efficient Markets for High Levels of Variable Renewable Energy*. Oxford, UK: Oxford Institute for Energy Studies.

Cronin, J., D. Fullerton, and S. Sexton. 2017. "Vertical and Horizontal Redistributions from a Carbon Tax and Rebate." Working paper. Cambridge, MA: National Bureau of Economic Research.

CRS (Congressional Research Service). 2010. *The Federal Government's Role in Electric Transmission Facility Siting*. Washington, DC.

CRS. 2018. *Water Resources Development Act of 2018 and America's Water Infrastructure Act of 2018: An Overview*. Washington, DC.

Cullenward, D., and D.G. Victor. 2020. *Making Climate Policy Work*. Cambridge, MA: Polity Press.

Cushing L, D. Blaustein-Rejto, M. Wander, M. Pastor, J. Sadd, A. Zhu, and R. Morello-Frosch. 2018. Carbon trading, co-pollutants, and environmental equity: Evidence from California's cap-and-trade program (2011–2015). *PLoS Medicine* 15(7).

C2ES (Center for Climate and Energy Solutions). 2019. "Carbon Tax Basics." https://www.c2es.org/content/carbon-tax-basics/.

C2ES. 2020a. *Getting to Zero: A U.S. Climate Agenda*. Arlington, VA.

C2ES. 2020b. "Federal Vehicle Standards." https://www.c2es.org/content/regulating-transportation-sector-carbon-emissions/.

C2ES (Center for Climate and Energy Solutions) and RAP (Regulatory Assistance Project). 2011. *Clean Energy Standards: State and Federal Policy Options and Implications*. Arlington, VA.

DiChristopher, T. 2020. "Banning' Natural Gas Is out; Electrifying Buildings Is In." S&P Global. https://www.spglobal.com/marketintelligence/en/news-insights/latest-news-headlines/banning-natural-gas-is-out-electrifying-buildings-is-in-59285807.

DOE (U.S. Department of Energy). 2015. *ARRA Grid Modernization Highlights: A Glimpse of the Future Grid through Recovery Act Funding*. Washington DC: Office of Energy Efficiency and Renewable Energy.

DOE. 2016. *Smart Grid Investment Grant Program Final Report*. Washington, DC: Office of Electricity Delivery and Energy Reliability.

DOE. 2017. *Quadrennial Energy Review. Transforming the Nation's Electricity System: The Second Installment of the QER*. Washington, DC.

DOE. 2018. *Strengthening Tribal Communities, Sustaining Future Generations*. Washington, DC: Department of Energy Office of Indian Energy.

DOE. 2019. "There are More Than 68,800 Electric Vehicle Charging Units in the United States." https://www.energy.gov/eere/vehicles/articles/fotw-1089-july-8-2019-there-are-more-68800-electric-vehicle-charging-units.

DOE. 2020a. "Frequently Answered Questions." https://www.eia.gov/tools/faqs/faq.php?id=86&t=1.

DOE. 2020b. "Resource Center | Building Energy Codes Program." https://www.energycodes.gov/resource-center.

DOE. 2020c. *FY 2021 Congressional Budget Request*. Volume 3 Part 2. Washington, DC.

DOE. n.d. "Buy American." https://www.energy.gov/gc/action-center-office-general-counsel/faqs-related-recovery-act/buy-american.

Doris E., J. Cochran, and M. Vorum. 2009. *Energy Efficiency Policy in the United States: Overview of Trends at Different Levels of Government*. Golden CO: National Renewable Energy Laboratory.

Driscoll, C.T., J.J. Buonocore, J.I. Levy, K.F. Lambert, D. Burtraw, S.B. Reid, et al. 2015. US power plant carbon standards and clean air and health co-benefits. *Nature Climate Change* 5: 535–540.

DSIRE. 2019. "Database of State Incentives for Renewables and Efficiency." https://www.dsireusa.org/resources/detailed-summary-maps/.

Eccles, R., and S. Klimenko. 2019. The investor revolution. *Harvard Business Review*. https://hbr.org/2019/05/the-investor-revolution.

EDF (Environmental Defense Fund). 2019. "Colorado Becomes First State in the Central U.S. to Adopt Zero Emission Vehicle Standards." https://www.edf.org/media/colorado-becomes-first-state-central-us-adopt-zero-emission-vehicle-standards.

EIA (Energy Information Administration). 2019. "Energy Power Annual." Table 10.10. October. https://www.eia.gov/electricity/annual/html/epa_10_10.html.

EIA. 2020. *U.S. Energy-Related Carbon Dioxide Emissions, 2019*. Paris, France.

EFI (Energy Futures Initiative). 2020. *An Action Plan for Carbon Capture and Storage in California: Opportunities, Challenges, and Solutions*. https://energyfuturesinitiative.org.

Ela, E., M. Milligan, A. Bloom, A. Botterud, A. Townsend, and T. Levin. 2014. *Evolution of Wholesale Electricity Market Design with Increasing Levels of Renewable Generation*. Golden, CO: National Renewable Energy Laboratory.

EPA (U.S. Environmental Protection Agency). 2009. *Interagency Report on International Competitiveness and Emission Leakage*. Washington, DC.

EPA. 2012. "Benchmarking and Energy Savings." https://www.energystar.gov/buildings/tools-and-resources/datatrends-benchmarking-and-energy-savings.

Eto, J. 2016. *Building Electric Transmission Lines: A Review of Recent Transmission Projects*. Berkeley, CA: Lawrence Berkeley National Laboratory.

Ex-Im Bank. *2019 Annual Report*. Washington, DC.

Farber, H., D. Herbst, I. Kuziemko, and S. Naidu. 2018. *Unions and Inequality Over the Twentieth Century: New Evidence from Survey Data*. Report no. w24587. Cambridge, MA: National Bureau of Economic Research.

FCC (Federal Communication Commission). 2019. *2019 Broadband Deployment Report*. Washington DC.

Fell, H., D. Burtraw, R. Morgenstern, and K. Palmer. 2011. Climate policy design with correlated uncertainties in offset supply and abatement cost. *Land Economics* 88(3): 589-611.

FERC (Federal Energy Regulatory Commission). 2011. "Demand Response Compensation in Organized Wholesale Energy Markets." Order No. 745. Washington, DC.

FERC. 2018. "Electric Storage Participation in Markets Operated by Regional Transmission Organizations and Independent System Operators." Order No. 841. https://ferc.gov/sites/default/files/2020-06/Order-841.pdf.

FERC. 2020. "Participation of Distributed Energy Resources." Order No. 2222. https://www.ferc.gov/sites/default/files/2020-09/E-1_0.pdf.

FHWA (Federal Highway Administration). 2020. "Alternative Fuel Corridors Program." https://www.fhwa.dot.gov/environment/alternative_fuel_corridors.

Fink, L. 2020. "BlackRock Capital, Letter to CEOs." https://www.blackrock.com/us/individual/larry-fink-ceo-letter.

Fischedick M., A. Roy, O. Edenhofer, R. Pichs-Madruga, Y. Sokona, E. Farahani, S. Kadner, et al. 2014. Industry. In *Climate Change 2014: Mitigation of Climate Change*. Contribution of Working Group III to the Fifth Assessment Report of the Intergovernmental Panel on Climate Change. Cambridge University Press, Cambridge, United Kingdom and New York, NY, USA.

Fischer, C. 2004. *Who Pays for Energy Efficiency Standards?* Washington, DC: Resources for the Future.

Fischer, C., and A.K. Fox. 2011. The role of trade and competitiveness measures in US climate policy. *American Economic Review* 101(3): 258–262.

Fischer, C., and R.G. Newell. 2008. Environmental and technology policies for climate mitigation. *Journal of Environmental Economics and Management* 55(2): 142–162.

Fischer, C., and W.A. Pizer. 2018. Horizontal equity effects in energy regulation. *Journal of the Association of Environmental and Resource Economists* 6(1): 209–237.

Fitzpatrick, R., J. McBride, J. Lovering, J. Freed, and T. Nordhaus. 2018. *Clean Energy Standards: How More States Can Become Climate Leaders—Third Way*. Washington, DC: Third Way and the Breakthrough Institute. http://thebreakthrough.org.

Fowlie, M., M. Greenstone, and C. Wolfram. 2018. Do energy efficiency investments deliver? Evidence from the Weatherization Assistance Program. *The Quarterly Journal of Economics* 133(3): 1597–1644.

Fox-Penner, P. 2020. *Power after Carbon: Building a Clean, Resilient Grid*. Cambridge, MA: Harvard University Press.

Frankel, J.A. 2008. "Addressing the Leakage/Competitiveness Issue in Climate Change Policy Proposals [with Comment]." Pp. 69–91 in *Brookings Trade Forum 2008/2009*. Washington, DC: Brookings Institution Press.

Friedmann, S.J. 2019. Congressional Testimony before the Committee on Energy and Commerce in the U.S. House of Representatives. September 18.

Fu, R., D. Feldman, and R. Margolis. 2018. *U.S. Solar Photovoltaic System Cost Benchmark: Q1 2018*. Golden, CO: National Renewable Energy Laboratory.

Gillingham, K., R. Newell, and K. Palmer. 2004. *Retrospective Examination of Demand-Side Energy Efficiency Policies*. Washington, DC: Resources for the Future.

Gillingham, K., R. Newell, and K. Palmer. 2006. Energy efficiency policies: A retrospective examination. *Annual Review of Environment and Resources* 31: 161–192.

Goldstein A., C. Doblinger, E. Baker, and L. Diaz Anadon. 2020 Patenting and business outcomes for cleantech startups funded by the Advanced Research Projects Agency-Energy. *Nature Energy* 5: 803–810.

Grab, D., and A. Shah. 2020. "California Can't Wait on All-Electric New Building Code." Rocky Mountain Institute. July 28. https://rmi.org/california-cant-wait-on-all-electric-new-building-code/.

Gramlich, R., and M. Hogan. 2019. *Wholesale Electricity Market Design for Rapid Decarbonization: A Decentralized Markets Approach*. San Francisco, CA: Energy Innovation.

Green, T., and C. Knittel. 2020. *Distributed Effects of Climate Policy: A Machine Learning Approach*. Roosevelt Project. Cambridge, MA: MIT's Center for Energy and Environmental Policy Research.

Greene, D.L., C.B. Sims, and M. Muratori. 2020. Two trillion gallons: Fuel savings from fuel economy improvements to US light-duty vehicles, 1975–2018. *Energy Policy* 142: 111–517.

Griffith-Jones, S., and J.A. Ocampo. *2018. The Future of National Development Banks. The Initiative for Policy Dialogue Series*. United Kingdom: Oxford University Press.

Griffith-Jones, S., S. Attridge, and M. Gouett. 2020. "Securing Climate Finance Through National Development Banks." Overseas Development Institution. http://www.stephanygj.net/papers/SecuringClimateFinanceThroughNational-DevelopmentBanksJan2020.pdf.

Haggerty, J.H., M.N. Haggerty, and R. Rasker. 2014. Uneven local benefits of renewable energy in the U.S. West: Property tax policy effects. *Western Economics Forum* 13(1): 1-16.

Hafstead, M., G.E. Metcalf, and R.C. Williams. 2016. Adding quantity certainty to a carbon tax through a tax adjustment mechanism for policy pre-commitment. *Harvard Environmental Law Review Forum* 41: 41–57.

Hafstead, M., and R.C. Williams III. 2020. Designing and evaluating a U.S. carbon tax adjustment mechanism to reduce emissions uncertainty. *Review of Environmental Economics and Policy* 14(1).

Hahn, R.W., and R.N. Stavins. 1992. Economic incentives for environmental protection: Integrating theory and practice. *The American Economic Review* 82(2): 464–468.

Hart, D.M. 2018. Beyond the technology pork barrel? An assessment of the Obama administration's energy demonstration projects. *Energy Policy* 119: 367–376.

Hausman, J.A., and P.L. Joskow. 1982. Evaluating the costs and benefits of appliance efficiency standards. *The American Economic Review* 72(2): 220–225.

Healey, P. 1998. Building institutional capacity through collaborative approaches to urban planning. *Environment Planning A: Economy and Space* 30(9): 1531–1546.

Hernandez-Cortes, D., and K.C. Meng. 2020. *Do Environmental Markets Cause Environmental Injustice? Evidence from California's Carbon Market*. Cambridge, MA: National Bureau of Economic Research.

Hibbard, P.J., and S.F Tierney. 2011. Carbon control and the economy: Economic impacts of RGGI's first three years. *The Electricity Journal* 24(10): 30-40.

Hibbard, P.J., S.F. Tierney, P. Darling, and S. Cullinan. 2018. An expanding carbon cap-and-trade regime? A decade of experience with RGGI charts a path forward. *The Electricity Journal* 31: 1-8.

Ho, M., R.D. Morgenstern, and J. Shih. 2008. *Impact of Carbon Price Policies on U.S. Industry*. Washington DC: Resources for the Future.

Hogan, W. 2014. Electricity market design and efficient pricing: Applications for New England and beyond. *The Electricity Journal* 27(7): 23-49.

Hogan, W. 2017. "Electricity Market Design Interactions of Multiple Markets." Presentation at RFF's Workshop on the Future of Power Markets in a Low Marginal Cost, September 14, 2017. https://media.rff.org/documents/170914_PowerMarkets_WilliamHogan.pdf.

Horowitz, J., J. Cronin, H. Hawkins, L. Konda, and A. Yuskavage. 2017. "Methodology for Analyzing a Carbon Tax." Working Paper 115. Washington, DC: Department of Treasury, Office of Tax Analysis.

Horowitz, J., R Igielnik, and R. Kochhar. 2020. *Most Americans Say There Is Too Much Economic Inequality in the U.S., but Fewer Than Half Call It a Top Priority*. Washington, DC: Pew Research Center.

Houde, S., and C.A. Spurlock. 2016. Minimum energy efficiency standards for appliances: Old and new economic rationales. *Economics of Energy and Environmental Policy* 5(2).

Hunt, J., and M. Gautheir-Loiselle. 2010. How much does immigration boost innovation? *American Economic Journal: Macroeconomics* 2(2): 31–56.

IEA. (International Energy Agency). 2011. *IEA Guide to Reporting Energy RD&D Budget/Expenditure Statistics*. Paris, France.

IEA. 2019. *World Energy Investment 2019*. Paris, France.

IEA. 2020. *Clean Energy Innovation*. Paris, France.

IMT (Institute for Market Transformation). 2020. "U.S. Building Benchmarking Policy, Landscape." https://www.buildingrating.org/graphic/us-building-benchmarking-policy-landscape.

Jaffe, A., S.R. Peterson, P.R. Portney, and R.N. Stavins. 1995. Environmental regulation and the competitiveness of U.S. manufacturing: What does the evidence tell us? *Journal of Economic Literature* 33(1): 132–163.

Jaffe, A., R. Newell, and R. Stavins. 2004. Economics of Energy Efficiency. *Encyclopedia of Energy* 2:79–90. Amsterdam: Elsevier.

Jaffe, A., R. Newell, and R. Stavins. 2005. A tale of two market failures: Technology and environmental policy. *Ecological Economics* 54(2-3): 164–174.

Jones, B., P. Philips, and C. Zabin. "The Link Between Good Jobs and a Low Carbon Future." University of California, Berkeley. https://laborcenter.berkeley.edu/pdf/2016/Link-Between-Good-Jobs-and-a-Low-Carbon-Future.pdf.

Jordan, G., and A. Link. 2018. *Evaluation of U.S. DOE Small Business Vouchers Pilot*. Washington, DC: Department of Energy Office of Energy Efficiency and Renewable Energy.

Joskow, P and R. Schmalensee. 2020. "MIT Energy Initiative's Energy Markets." Podcast Episode #14. http://energy.mit.edu/podcast/electricity-markets/.

Jungclaus, M., C. Carmichael, C. McClurg, M. Simmons, R. Smidt, K.P. Hydras, S. Conger, et al. 2017. Deep energy retrofits in federal buildings: The value, funding models, and best practices. *ASHRAE Transactions* 124(1).

Kaufman, N., A.R. Barron, W. Krawczyk, P. Marsters, and H. McJeon. 2020. A near-term to net zero alternative to the social cost of carbon for setting carbon prices. *Nature Climate Change* 10: 1010–1014.

Kortum, S., and D. Weisbach. 2017. The design of border adjustments for carbon pricing. *National Tax Journal* 70(2): 421–446.

Krishnaswami, A., and J. Higdon. 2020. *A Progressive Climate Innovation Agenda*. Data for Progress. https://www.dataforprogress.org.

Larsen, J., N. Kaufman, P. Marsters, W. Herndon, H. Kolus, and B. King. 2020. *Expanding the Reach of a Carbon Tax: Emissions Impacts of Pricing Combined with Additional Climate Actions*. New York, NY: Columbia SIPA Center on Global Energy Policy.

Larson, E., C. Greig, J. Jenkins, E. Mayfield, A. Pascale, C. Zhang, S. Pacala, et al. 2020. *Net-Zero America by 2050: Potential Pathways, Deployments, and Impacts*. Princeton, NJ: Princeton University.

Larson J., S. Mohan, P. Marsters, and W. Herdon. 2018. *Energy and Environmental Implications of a Carbon Tax in the United States*. New York, NY: Columbia SIPA Center on Global Energy Policy and Rhodium Group.

Levin, B. 2019. "A Broadband Agenda for the (Eventual) Infrastructure Bill." Brookings Institution. https://www.brookings.edu/blog/the- avenue/2019/03/19/a-broadband-agenda-for-the-eventual-infrastructure-bill/.

Litterman, R. 2020a. "Financial Regulation and Climate Risk Management." Interview with Robert Litterman. Harvard Business School's "Climate Rising" podcast. https://www.hbs.edu/environment/podcast/Pages/podcast-details.aspx?episode=15116819.

Litterman, R. 2020b. "Pricing Climate Risk in the Markets, with Robert Litterman." Resources Radio. https://www.resourcesmag.org/resources-radio/pricing-climate-risk-markets-robert-litterman/.

Luna-Martinez, J., and C.L. Vicente. 2012. *Global Survey of Development Banks*. Washington, DC: The World Bank.

Lyubich, E. 2020. *The Race Gap in Residential Energy Expenditures*. Berkeley, CA: Berkeley Haas.

MacDonald, A., C. Clack, A. Alexander, A. Dunbar, J. Wilczek, and Y. Xie. 2016. Future cost-competitive electricity systems and their impact on US CO_2 emissions. *Nature Climate Change* 6: 526-531.

Margolies, J. 2020. "'All-Electric' Movement Picks Up Speed, Catching Some Off Guard." *The New York Times*. February 4. https://www.nytimes.com/2020/02/04/business/all-electric-green-development.html.

Meng, T., D. Hsu, and A. Han. 2016. *Measuring Energy Savings from Benchmarking Policies in New York City, ACEEE Summer Study on Energy Efficiency in Buildings*. Washington, DC: American Council for an Energy-Efficient Economy.

Metcalf, G.E. 2008. Using tax expenditures to achieve energy policy goals. *American Economic Review* 98(2): 90–94.

Metcalf, G.E. 2009. Market-based policy options to control U.S. greenhouse gas emissions. *Journal of Economic Perspectives* 23(2): 5–27.

Metcalf, G.E. 2020. An emissions assurance mechanism: Adding environmental certainty to a U.S. carbon tax. *Review of Environmental Economics and Policy* 14(1).

Moran, D., A. Hasanbeigi, and C. Springer. 2018. *The Carbon Loophole in Climate Policy*. KGM & Associates, ClimateWorks Foundation and Global Efficiency Energy Intelligence.

Morgan, M. 2019. "Buy American vs. Buy America: A Simple Guide to Successfully Navigating the Differences." https://www.mbpce.com/blog/buy-american-vs-buy-america-a-simple-guide-to-successfully-navigating-the-differences/.

Morrison, T.H. 2014. Developing a regional governance index: The institutional potential of rural regions. *Journal of Rural Studies* 35: 101–111.

Mufson, S. 2020. "The Fastest Way to Cut Carbon Emissions Is a "Fee" and a Dividend, Top Leaders Say." *The Washington Post*. https://www.washingtonpost.com/climate-environment/the-fastest-way-to-cut-carbon-emissions-is-a-fee-and-a-rebate-top-leaders-say/2020/02/13/b63b766c-4cfc-11ea-bf44-f5043eb3918a_story.html.

Muro, M., A. Tomer, R. Shivaram, and J. Kane. 2019. *Advancing Inclusion through Clean Energy Jobs*. Washington, DC: Brookings Institute.

Murray, A.G., and B.F. Mills. 2014. The impact of Low-Income Home Energy Assistance Program participation on household energy insecurity. *Contemporary Economic Policy* 32(4): 811–825.

Murray, B.C., W.A. Pizer, and C. Reichert. 2017. Increasing emissions certainty under a carbon tax. *Harvard Law Review Forum* 41: 13–27.

Myslikova, Z., and K.S. Gallagher. 2020. Mission innovation is mission critical. *Nature Energy* 5: 732–734.

Nadel, S., N. Elliot, and T. Langer. 2015. *Energy Efficiency in the United States: 35 Years and Counting*. Washington, DC: American Council for an Energy-Efficient Economy.

NASEM (National Academies of Sciences, Engineering, and Medicine). 2017a. *An Assessment of ARPA-E*. Washington, DC: The National Academies Press.

NASEM. 2017b. *Enhancing the Resilience of the Nation's Electricity System*. Washington, DC: The National Academies Press.

NASEO (National Association of State Energy Officials) and EFI (Energy Futures Initiative). 2019. *The 2019 U.S. Energy and Employment Report 2019*. https://www.usenergyjobs.org.

NASEO and EFI. 2020. *2020 U.S. Energy and Employment Report*. Washington, DC.

Nemet, G.F., V. Zipperer, and M. Kraus. 2018. The valley of death, the technology pork barrel, and public support for large demonstration projects. *Energy Policy* 119: 154–167.

Newbery, D.M. 2008. Climate change policy and its effect on market power in the gas market. *Journal of the European Economic Association* 6(4): 727–751.

NGA (National Governors Association). 2019. "Clean Energy Toolkit: Growing Electrification." https://www.nga.org/wp-content/uploads/2019/11/NGA_CleanEnergy_Toolkit_Growing_Electrification.pdf.

NPR. 2019. "For Many Navajos, Getting Hooked Up To the Power Grid Can Be Life-Changing." https://www.npr.org/sections/health-shots/2019/05/29/726615238/for-many-navajos-getting-hooked-up-to-the-power-grid-can-be-life-changing.

NPR, Robert Wood Johnson Foundation, and Harvard T.H. Chan School of Public Health. 2019. *Life in Rural America Part II*. https://www.rwjf.org/en/library/research/2019/05/life-in-rural-america—part-ii.html.

NRC (National Research Council). 2006. *State and Federal Standards for Mobile-Source Emissions*. Washington, DC: The National Academies Press.

NREL (National Renewable Energy Laboratory). 2017. "Green Banks." https://www.nrel.gov/state-local-tribal/basics-green-banks.html.

NSF (National Science Foundation). 2020. *Science and Engineering Indicators*. Washington, DC.

NYC (New York City). 2019. "Local Laws of the City of New York for the Year 2019, No. 97." https://www1.nyc.gov/assets/buildings/local_laws/ll97of2019.pdf.

Office of the Governor. 2020. "Governor Newsom Announces California Will Phase Out Gasoline-Powered Cars and Drastically Reduce Demand for Fossil Fuel in California's Fight Against Climate Change." https://www.gov.ca.gov/2020/09/23/governor-newsom-announces-california-will-phase-out-gasoline-powered-cars-drastically-reduce-demand-for-fossil-fuel-in-californias-fight-against-climate-change/.

Olson, E. 2020. *Federal Investment in EV Charging Infrastructure for Economic Recovery, Climate, and Public Health*. The Breakthrough Institute. http://thebreakthrough.org.

Omarova, S. 2020. *The Climate Case for a National Investment Authority*. Data for Progress. https://www.dataforprogress.org.

Palmer, K., and M. Walls. 2015. *Does Information Provision Shrink the Energy Efficiency Gap?* Washington, DC: Resources for the Future.

Palmer, K., A. Paul, and A. Keyes. 2018. *Hanging Baselines, Shifting Margins: How Predicted Impacts of Pricing Carbon in the Electricity Sector Have Evolved over Time*. Washington, DC: Resources for the Future.

Partridge, A., and B. Steigauf. 2020. *Minnesota's Power Plant Communities: An Uncertain Future*. Minneapolis-St. Paul, MN: Center for Energy and Environment.

PCAST (President's Council of Advisors on Science and Technology). 2010. *Report to the President on Accelerating the Pace of Change in Energy Technologies Through an Integrated Federal Energy Policy*. Washington, DC.

Perrin, A. 2019. "Digital Gap Between Rural and Nonrural America Persists." Pew Research Center. https://www.pewresearch.org/fact-tank/2019/05/31/digital-gap-between-rural-and-nonrural-america-persists/.

Pierpont, B., and D. Nelson. 2017. *Markets for Low Carbon, Low Cost Electricity Systems*. San Francisco, CA: Climate Policy Initiative.

Pizer, W.A, and S. Sexton. 2019. The distributional impacts of energy taxes. *Review of Environmental Economics and Policy* 13(1): 104–123.

Pomerleau, K., and E. Asen, E. 2019. *Carbon Tax and Revenue Recycling: Revenue, Economic, and Distributional Implications*. Washington, DC: Tax Foundation.

Ramani, V. 2020. *Addressing Climate Risk as a Systemic Risk: A Call to Action for U.S. Financial Regulators*. Arlington, VA: The CERES Accelerator for Sustainable Capital Markets.

Rausch, S., G.E. Metcalf, and J.M. Reilly. 2011. Distributional impacts of carbon pricing: A general equilibrium approach with micro-data for households. *Energy Economics* 33(1): 20-33.

Reed, L., M.G. Morgan, P. Vaishnay, and D.E. Armanios. 2019. Converting existing transmission corridors to HVDC is an overlooked option for increasing transmission capacity. *Proceedings of the National Academy of Sciences U.S.A.* 116(28): 13879-13884.

Reed, L., M. Dworkin, P. Vaishnav, and M.G. Morgan. 2020. Expanding transmission capacity: Examples of regulatory paths for five alternative strategies. *The Electricity Journal* 33(6).

RGGI, Inc. 2020. "Investments of Proceeds." https://www.rggi.org/investments/proceeds-investments.
Rosenbloom, D., J. Markard, F.W. Geels, and L. Fuenfschilling. 2020. Opinion: Why carbon pricing is not sufficient to mitigate climate change—and how "sustainability transition policy" can help. *Proceedings of the National Academy of Sciences U.S.A.* 117(16): 8664–8668.
Rozansky, R., and D.M. Hart. 2020. *More and Better: Building and Managing a Federal Energy Demonstration Project Portfolio*. Washington, DC: Information Technology and Innovation Foundation.
Ruckelshaus, C., and S. Leberstein. 2014. *Manufacturing Low Pay: Declining Wages in the Jobs That Built America's Middle Class*. New York, NY: National Employment Law Project.
Schwabe, P. 2016. *Solar Energy Prospecting in Remote Alaska*. Washington DC: U.S. Department of Energy, Office of Indian Energy.
Scott, R.E. 2020. "We Can Reshore Manufacturing Jobs, but Trump Hasn't Done It." Economic Policy Institute. https://www.epi.org/publication/reshoring-manufacturing-jobs/.
Shah, T., and A. Krishnaswami. 2019. *Transforming the U.S. Department of Energy in Response to the Climate Crisis*. New York, NY: Natural Resources Defense Council.
Shierholz, H. 2019. *Working People Have Been Thwarted in Their Efforts to Bargain for Better Wages by Attacks on Unions*. Washington, DC: Economic Policy Institute.
Sivaram, V., C. Cunliff, D. Hart, J. Friedmann, and D. Sandalow. 2020. *Energizing America: A Roadmap to Launch a National Energy Innovation Mission*. New York, NY: Columbia Center on Global Environmental Policy.
Stavins, R. 2009. "The Wonderful Politics of Cap-and-Trade: A Closer Look at Waxman-Markey." An Economic View of the Environment. May 27. http://www.robertstavinsblog.org/2009/05/27/the-wonderful-politics-of-cap-and-trade-a-closer-look-at-waxman-markey/.
Stepp, M., S. Pool, J. Spencer, and N. Loris. 2013. *Turning the Page: Reimagining the National Labs in the 21st Century Innovation Economy*. Washington, DC: Information Technology and Innovation Foundation.
Stehly, T., and P. Beiter. 2019. *2018 Cost of Wind Energy Review*. Golden, CO: National Renewable Energy Laboratory.
Swanstrom, D., and M.M. Jolivert. 2009. DOE transmission corridor designations and FERC backstop siting authority: Has the Energy Policy Act of 2005 succeeded in stimulating the development of new transmission facilities? *Energy Law Journal* 30: 415-466.
TCFD (Task Force on Climate-related Financial Disclosures). 2017. *Final TCFD Recommendations Report: Executive Summary*. June. https://www.fsb-tcfd.org/publications/.
Terman, J.N. 2018. Helping third-party implementers meet performance obligations: A multi-level examination of the Weatherization Assistance Program. *Public Administration Quarterly* 42(3).
Tierney, S., and P.H. Hibbard. 2019. *Clean Energy in New York State: The Role and Economic Impacts of a Carbon Price in NYISO's Wholesale Electricity Markets*. Boston, MA: Analysis Group.
Tonn B., D. Carroll, E. Rose, B. Hawkins, S. Pigg, D. Bausch, G. Dalhoff, M. Blasnik, J. Eisenberg, C. Cowan, and B. Conlon. 2015. *Weatherization Works II—Summary of Findings from the ARRA Period Evaluation of the U.S. Department of Energy's Weatherization Assistance Program*. Oak Ridge, TN: Oak Ridge National Laboratory.
Tonn, B., E. Rose, and B. Hawkins. 2018. Evaluation of the U.S. Department of Energy's Weatherization Assistance Program: Impact results. *Energy Policy* 118: 279–290.
Traut, E.J., T.C. Cherng, C. Hendrickson, and J.J. Michalek. 2013. US residential charging potential for electric vehicles. *Transportation Research Part D: Transport and Environment* 25: 139–145.
TRB (Transportation Research Board) and NRC (National Research Council). 2015. *Overcoming Barriers to Deployment of Plug-in Electric Vehicles*. Washington, DC: The National Academies Press.
U.N. Development Programme. 2019. *Human Development Report 2019*. New York, NY.
Urbanek, L. 2017. "Changes to the Standards Program: More Harm than Good?" https://www.nrdc.org/experts/lauren-urbanek/changes-standards-program-more-harm-good.
U.S. Congress. House. 2020a. "E&C Leaders Release Framework of the CLEAN Future Act, a Bold New Plan to Achieve a 100 Percent Clean Economy by 2050." Press release. House Committee on Energy and Commerce. https://energycommerce.house.gov/newsroom/press-releases/ec-leaders-release-framework-of-the-clean-future-act-a-bold-new-plan-to.

U.S. Congress. House. 2020b. *Solving the Climate Crisis: The Congressional Action Plan for a Clean Energy Economy and a Healthy, Resilient and Just America*. Majority Staff Report. Select Committee on the Climate Crisis. Washington, DC.

U.S. Congress. House. 2020c. "Summary of the Climate Leadership and Environmental Action for Our Nation's (CLEAN) Future Act." House Committee on Energy and Commerce. https://energycommerce.house.gov/sites/democrats.energycommerce.house.gov/files/documents/Section-by-Section%20of%20CLEAN%20Future%20Act%20.pdf.

U.S. Congress. Senate. 2019. "FY2020 Energy and Water Development Appropriations Bill Advanced by Full Committee." News. Committee on Appropriations. https://www.appropriations.senate.gov/news/fy2020-energy-and-water-development-appropriations-bill-advanced-by-full-committee.

U.S. Congress. Senate. 2020. "Murkowski, Manchin Introduce American Energy Innovation Act." Republican News. Committee on Energy and Natural Resources. February 27. https://www.energy.senate.gov/2020/2/murkowski-manchin-introduce-american-energy-innovation-act.

USG (U.S. Government). 2017. "View Rule." https://www.reginfo.gov/public/do/eAgendaViewRule?pubId=201704&RIN=1904-AE11.

Vaughan, E., and J. Turner. 2013. *The Value and Impact of Building Codes*. Washington, DC: Environmental and Energy Study Institute.

Vizcarra, H. 2020. *The Reasonable Investor and Climate-Related Information: Changing Expectations for Financial Disclosures*. Washington, DC: Environmental Law Institute.

Voos, P. 2009. "How Unions Can Help Restore the Middle Class." Economic Policy Institute. https://www.epi.org/publication/how_unions_can_help_restore_the_middle_class/.

Wallace, M., L. Goudarzi, K. Callahan, and R. Wallace. 2015. *A Review of the CO_2 Pipeline Infrastructure in the U.S.* Washington, DC: National Energy Technology Laboratory.

Weitzman, M.L. 1974. Prices vs. quantities. *Review of Economic Studies* 41(4): 477–491.

Weitzman, M. 2009. On modeling and interpreting the economics of catastrophic climate change. *The Review of Economics and Statistics* 91(1): 1-19.

Wigley, T., R. Richels, and J. Edmonds. 1996. Economic and environmental choices in the stabilization of atmospheric CO_2 concentrations. *Nature* 379: 240–243.

Williams, R.C., H. Gordon, D. Burtraw, J.C. Carbone, and R.D. Morgenstern. 2015. The initial incidence of a carbon tax across income groups. *National Tax Journal* 68(1): 195–213.

Wiseman, H. 2011. Expanding regional renewable governance. *Harvard Environmental Law Review* 35(2): 477-540.

World Bank. 2020. "Stock Market Capitalization to GDP for United States [DDDM01USA156NWDB]." https://data.worldbank.org/indicator/CM.MKT.TRAD.GD.ZS.

Zaidi, A. 2020. "Mandates for Action: Corporate Governance Meets Climate Change." *Stanford Law Review Online*. https://www.stanfordlawreview.org/online/new-mandates-for-action/.

APPENDIX A

Committee Biographical Information

STEPHEN W. PACALA, *Chair,* is the Frederick D. Petrie Professor of Ecology and Evolutionary Biology at Princeton University. Dr. Pacala directs the Carbon Mitigation Initiative, an effort to develop solutions to the greenhouse warming problem. He is also a founder and chair of the board of Climate Central, a nonprofit media organization focusing on climate change. Dr. Pacala chaired the Committee on Carbon Dioxide Removal and Sequestration of the National Academies of Sciences, Engineering, and Medicine, which released its report in 2018. His research covers a wide variety of ecological and mathematical topics with an emphasis on interactions between greenhouse gases (GHGs), climate, and the biosphere. Dr. Pacala received an undergraduate degree from Dartmouth College in 1978 and a Ph.D. in biology from Stanford University in 1982. He serves on the boards of the Environmental Defense Fund and Hamilton Insurance Group. Among his many honors are the David Starr Jordan Prize and the George Mercer Award of the Ecological Society of America. Dr. Pacala is a member of the National Academy of Sciences and the American Academy of Arts and Sciences.

COLIN CUNLIFF is a senior policy analyst in clean energy innovation policy at the Information Technology and Innovation Foundation, a nonpartisan think tank and research organization. Dr. Cunliff previously worked at the U.S. Department of Energy (DOE) on climate mitigation and energy sector resilience. At DOE, he contributed to the *Quadrennial Energy Review: Transforming the Nation's Electricity System*, a national roadmap to modernize the U.S. electricity system. Prior to that, Dr. Cunliff served as the American Institute of Physics (AIP) Congressional Fellow in the office of Senator Dianne Feinstein (D-CA), where he was a staff science adviser on energy, climate, and transportation. He holds a Ph.D. in physics from the University of California, Davis, and a bachelor's of science in physics and mathematics from the University of Texas, Austin.

DANIELLE DEANE-RYAN has devoted her career to her passion for forging equitable climate crisis solutions. Ms. Deane-Ryan is an independent consultant serving as a senior advisor to foundations including the Donors of Color Network and the Libra Foundation. Previously, she directed the Inclusive Clean Economy Program at the Nathan Cummings Foundation, supporting collaborations that catalyzed world-leading climate and equity policies. Ms. Deane-Ryan served in the Obama administration as senior advisor for external affairs and acting director for stakeholder engagement at

APPENDIX A

the DOE Office of Energy Efficiency and Renewable Energy. She is a co-author of the 2019 Clean Energy States Alliance Report *Solar with Justice: Strategies for Powering Up Under-Resourced Communities and Growing an Inclusive Solar Market*. Prior, Ms. Deane-Ryan was the founding executive director of Green 2.0 and a principal of the Raben Group; launched the New Constituencies for Environmental Program at the Hewlett Foundation; and managed the Commission to Engage African Americans on Energy, Climate Change, and the Environment at the Joint Center for Political and Economic Studies. Ms. Deane-Ryan is on the boards of the Clean Energy States Alliance and Resource Media. She holds an M.Sc. from the London School of Economics in environment and development and a B.A. from Williams College in political economy with an environmental studies concentration. Williams College awarded its Bicentennial Medal in 2019 to Ms. Deane-Ryan for her contributions to the environmental justice field.

KELLY SIMS GALLAGHER is a professor of energy and environmental policy at the Fletcher School, Tufts University. Dr. Gallagher directs the Climate Policy Lab and the Center for International Environment and Resource Policy at Fletcher. From June 2014 to September 2015, she served in the Obama administration as a senior policy advisor in the White House Office of Science and Technology Policy, and as senior China advisor in the Special Envoy for Climate Change Office at the U.S. State Department. Dr. Gallagher is a member of the board of the Belfer Center for Science and International Affairs at Harvard University. She is a member of the executive committee of the Tyler Prize for Environmental Achievement, she serves on the board of the Energy Foundation, and she is a member of the Council on Foreign Relations. Broadly, Dr. Gallagher focuses on energy innovation and climate policy. She specializes in how policy spurs the development and deployment of cleaner and more efficient energy technologies, domestically and internationally. Dr. Gallagher is the author of *Titans of the Climate* (2018); *The Global Diffusion of Clean Energy Technologies: Lessons from China* (2014); *China Shifts Gears: Automakers, Oil, Pollution, and Development* (2006); and dozens of other publications.

JULIA HAGGERTY is an associate professor of geography in the Department of Earth Sciences at Montana State University (MSU), where she holds a joint appointment in the Montana Institute on Ecosystems. Dr. Haggerty received her B.A. in liberal arts from Colorado College and her Ph.D. in history from the University of Colorado. An award-winning teacher, Dr. Haggerty teaches courses in human, economic, and energy resource geography at MSU. She also leads the Resources and Communities Research Group in studying the ways rural communities respond to shifting economic and policy trajectories, especially as they involve natural resources. Dr. Haggerty has expertise in diverse rural geographies, including those shaped by energy development, extractive industries, ranching and agriculture, and amenity development and conservation. Partnerships and collaboration with diverse stakeholders are central to her approach. Prior to joining MSU, Dr. Haggerty was a postdoctoral fellow at the University of Otago in New Zealand (2005–2007) and a policy analyst with Headwaters Economics in Bozeman, Montana (2008–2013). She speaks frequently to public audiences about her research and has served on a number of boards and advisory committees from local to international scales.

CHRISTOPHER T. HENDRICKSON is the Hamerschlag University Professor of Engineering Emeritus, director of the Traffic 21 Institute at Carnegie Mellon University, and editor-in-chief of the *ASCE Journal of Transportation Engineering Part A (Systems)*. Dr. Hendrickson's research, teaching, and consulting are in the general area of engineering planning and management, including transportation systems, design for the environment, system performance, construction project management, finance, and computer applications. Central themes in his work are a systems-wide perspective and a balance of engineering and management considerations. He has co-authored eight books and published numerous articles in the professional literature. Dr. Hendrickson has been the recipient of the Council of University Transportation Centers Lifetime Achievement Award (2020), the ARTBA Steinburg Award (2019), the Faculty Award of the Carnegie Mellon Alumni Association (2009), the Turner Lecture Award of the American Society of Civil Engineers (2002), and the Fenves Systems Research Award from the Institute of Complex Engineering Systems (2002). He is a member of the National Academy of Engineering and the National Academy of Construction (2014), fellow of the American Association for the Advancement of Science (2007), a distinguished member of the American Society of Civil Engineers (2007), chair of the Transportation Research Board Division Committee, and former member of the Transportation Research Board (2004). Dr. Hendrickson earned bachelor's and M.S. degrees from Stanford University, an M.Phil. degree in economics from Oxford University, and a Ph.D. from the Massachusetts Institute of Technology (MIT).

JESSE D. JENKINS is an assistant professor at Princeton University with a joint appointment in the Department of Mechanical and Aerospace Engineering and the Andlinger Center for Energy and the Environment. Dr. Jenkins is an energy systems engineer with a focus on the rapidly evolving electricity sector, including the transition to zero-carbon resources, the proliferation of distributed energy resources, and the role of electricity in economy-wide decarbonization. His research focuses on improving and applying optimization-based energy systems models to evaluate low-carbon energy technologies, policy options, and robust decisions under deep uncertainty. Dr. Jenkins completed a Ph.D. in engineering systems and an M.S. in technology and policy at MIT and a B.S. in computer and information science at the University of Oregon. He worked previously as a postdoctoral environmental fellow at the Harvard Kennedy

School and the Harvard University Center for the Environment, a researcher at the MIT Energy Initiative, the director of Energy and Climate Policy at the Breakthrough Institute, and a policy and research associate at Renewable Northwest.

ROXANNE JOHNSON established and currently directs the Research Department at the BlueGreen Alliance (BGA), a national coalition of labor unions and environmental groups working to build a stronger, fairer economy. In Ms. Johnson's current role, she leads BGA's research efforts to understand job creation opportunities in the clean economy. Her team is responsible for conducting manufacturing and policy research in industries such as wind and solar energy, energy efficiency, advanced vehicles, and infrastructure. Her previous work at the Great Plains Institute focused on communicating model results showing potential impacts of energy and transportation policy. Ms. Johnson earned a B.S. in mathematics and environmental studies from Northland College in Ashland, Wisconsin. She also earned an M.S. in science, technology, and environmental policy from the Humphrey School of Public Affairs in Minneapolis, Minnesota.

TIMOTHY C. LIEUWEN serves as executive director of the Strategic Energy Institute at the Georgia Institute of Technology. Dr. Lieuwen is also a Regents' Professor and the David S. Lewis Jr. Chair in the School of Aerospace Engineering. He is also founder and chief technology officer of TurbineLogic, an analytics firm working in the gas turbine industry. Dr. Lieuwen is an international authority on gas turbine technologies, both from a research and development perspective and from a field/operational perspective. He has authored or edited four books, including the textbook *Unsteady Combustor Physics*. He has authored more than 350 publications and received four patents, all of which are licensed to the gas turbine industry. Dr. Lieuwen is editor-in-chief of the American Institute of Aeronautics and Astronautics (AIAA) *Progress* book series. He is also past chair of the Combustion, Fuels, and Emissions Technical Committee of the American Society of Mechanical Engineers (ASME) and has served as associate editor of *Combustion Science and Technology, Proceedings of the Combustion Institute*, and *AIAA Journal of Propulsion and Power*. Dr. Lieuwen is a fellow of ASME and AIAA, and a recipient of the AIAA Lawrence Sperry Award, ASME George Westinghouse Gold Medal, National Science Foundation (NSF) CAREER Award, and various best paper awards. Board positions include appointment by the secretary of energy to the National Petroleum Counsel, board of governors of Oak Ridge National Laboratory, and board member of the ASME International Gas Turbine Institute. Dr. Lieuwen has also served on a variety of federal review and advisory committees. He holds a Ph.D. in mechanical engineering from Georgia Tech. He has served on the National Academies Review of NASA Test Flight Capabilities and the decadal survey of aeronautics committees.

VIVIAN LOFTNESS is a university professor and former head of the School of Architecture at Carnegie Mellon University. Ms. Loftness is an internationally renowned researcher, author, and educator with more than 30 years of focus on environmental design and sustainability, advanced building systems integration, climate and regionalism in architecture, and design for performance in the workplace of the future. She has served on 10 National Academies panels and on the Board on Infrastructure and the Constructed Environment and has given four congressional testimonies on sustainability. Ms. Loftness is a recipient of the National Educator Honor Award from the American Institute of Architecture Students and the Sacred Tree Award from the U.S. Green Building Council (USGBC). She received her B.S. and M.S. in architecture from MIT and served on the National Boards of the USGBC, American Institute of Architects (AIA) Committee on the Environment, Green Building Alliance, Turner Sustainability, and the Global Assurance Group of the World Business Council for Sustainable Development. Ms. Loftness is a registered architect and a fellow of the AIA.

CLARK A. MILLER is a professor and director of the Center for Energy and Society at Arizona State University (ASU). Dr. Miller leads sustainability research for the Quantum Energy and Sustainable Solar Technologies Engineering Research Center. He also serves as a member of the steering committee of LightWorks, ASU's university-wide sustainable energy initiative. Dr. Miller's current research focuses on the human and social dimensions of energy transitions, including the social value of distributed renewable energy systems, strategies for addressing poverty and inequality through energy innovation, the organization of urban and regional energy transitions, and the design and governance of solar energy futures. He is an author or editor of eight books: *The Weight of Light* (2019); *Designing Knowledge* (2018); *The Handbook of Science and Technology Studies* (2016); *The Practices of Global Ethics* (2015); *Science and Democracy* (2015); *Nanotechnology, the Brain, and the Future* (2013); *Arizona's Energy Future* (2011); and *Changing the Atmosphere* (2001). Dr. Miller has published extensively in the fields of energy policy, science and technology policy, the role of science in democratic governance and international relations, the governance of emerging technologies, and the design of knowledge systems for improved decision making. He holds a Ph.D. in electrical engineering from Cornell University.

WILLIAM A. PIZER is the Susan B. King Professor and senior associate dean for faculty and research at the Sanford School of Public Policy and faculty fellow at the Nicholas Institute for Environmental Policy Solutions, both at Duke University. Dr. Pizer is also a university fellow at Resources for the Future and a research associate at the National Bureau of Economic Research. His current research examines how we value the future benefits of climate change mitigation, how environmental regulation and climate policy can affect production costs and competitiveness, and how the design of

APPENDIX A

market-based environmental policies can address the needs of different stakeholders. Dr. Pizer has been actively involved in the creation of an environmental program at Duke Kunshan University in China, a collaborative venture between Duke University, Wuhan University, and the city of Kunshan. Before coming to Duke, he was deputy assistant secretary for environment and energy at the U.S. Department of the Treasury from 2008 to 2011, overseeing Treasury's role in the domestic and international environment and energy agenda of the United States. Prior to that, he was a researcher at Resources for the Future for more than a decade. Dr. Pizer has written more than 50 peer-reviewed publications, books, and articles, and holds a Ph.D. and an M.A. in economics from Harvard University and a B.S. in physics from the University of North Carolina, Chapel Hill.

VARUN RAI is an associate professor in the LBJ School of Public Affairs and in the Department of Mechanical Engineering at the University of Texas, Austin, where he directs the Energy Systems Transformation Research Group (also known as the "Rai Group"). Dr. Rai's interdisciplinary research—delving into issues at the interface of energy systems, complex systems, decision science, and public policy—focuses on studying how the interactions between the underlying social, behavioral, economic, technological, and institutional components of the energy system impact the diffusion of energy technologies. Over the past 15 years, his research has applied various analytical lenses to the study of technologies and policies in carbon capture and sequestration, fuel cells, oil and gas, plug-in hybrid vehicles, and solar photovoltaics. Dr. Rai has presented at several important forums, including the U.S. Senate Briefings, Global Intelligent Utility Network Coalition, and Global Economic Symposium, and his research group's work has been discussed in the *New York Times*, *Wall Street Journal*, *Washington Post*, and *Bloomberg News*, among other venues. He was a Global Economic Fellow in 2009 and holds the Elspeth Rostow Centennial Fellowship at the LBJ School. From 2013 to 2015, Dr. Rai was a commissioner for the vertically integrated electric utility Austin Energy. In 2016, the Association for Public Policy Analysis and Management awarded him the David N. Kershaw Award and Prize, which "was established to honor persons who, at under the age of 40, have made a distinguished contribution to the field of public policy analysis and management." He received the Eyes of Texas Excellence Award, also in 2016, for making "noteworthy contributions to the UT community." Dr. Rai has been the associate dean for research at the LBJ School since 2017. He received his Ph.D. and M.S. in mechanical engineering from Stanford University and a bachelor's degree in mechanical engineering from the Indian Institute of Technology (IIT), Kharagpur.

ED RIGHTOR is the director of the Industrial Program at the American Council for an Energy-Efficient Economy (ACEEE). In this role, Dr. Rightor develops and leads the strategic vision for the industrial sector, shapes the research and policy agenda, and convenes stakeholders to accelerate energy efficiency and reductions of GHGs. Prior to joining ACEEE, Dr. Rightor held several leadership roles at Dow Chemical during his 30-year career. Through 2017, he served as the director of strategic projects in Dow's Environmental Technology Center. In this role, he worked with Dow businesses, operations, and corporate groups to reduce air emissions, waste, freshwater intake, and energy use. Dr. Rightor also served as the facilitator of Dow's Corporate Water Strategy Team, led teams to establish and pursue Dow's 2025 Sustainability Goals, including the first-ever water goal. Working across global industrial associations, he spearheaded a roadmap for the chemical industry on paths to reduce energy and GHG emissions. In prior roles, Dr. Rightor developed GHG and energy reduction options across Dow's global operations and pursued project funding and implementation. Earlier, he started a new market-facing business in the energy sector, led cross-functional teams to optimize processes (six sigma), pioneered technology that led to new materials development, and led teams to troubleshoot production challenges. Dr. Rightor earned a doctorate in chemistry from Michigan State University and a bachelor's of science in chemistry from Marietta College.

ESTHER TAKEUCHI is a professor at Stony Brook University and a chief scientist at the Brookhaven National Laboratory. Dr. Takeuchi is an energy storage expert who led efforts to invent and refine the lifesaving lithium/silver vanadium oxide (Li/SVO) battery technology, utilized in the majority of today's implantable cardioverter defibrillators (ICDs). Dr. Takeuchi's work was conducted during 22 years at Greatbatch, Inc., a major supplier of pacemaker and ICD batteries. ICD batteries have high energy density with the ability to support intermittent high-power pulses. In addition, they have a long life, are safe, and are durable. In Dr. Takeuchi's innovation, the cathodes employ two metals, silver and vanadium, rather than just one, allowing for more energy. In addition, the Li/SVO chemistry lets the ICD monitor the level of discharge, allowing it to predict end of service in a reliable manner. Today, more than 300,000 ICDs are implanted every year. Dr. Takeuchi received her B.A. from the University of Pennsylvania and her Ph.D. from the Ohio State University. She joined Greatbatch in 1984, and in 2007, she joined the State University of New York, Buffalo. Dr. Takeuchi is a member of the National Academy of Engineering, has received more than 140 U.S. patents, and is the recipient of the 2008 National Medal of Technology and Innovation.

SUSAN F. TIERNEY, a senior advisor at Analysis Group, is an expert on energy economics, regulation, and policy, particularly in the electric and gas industries. Dr. Tierney consults to businesses, government agencies, foundations, tribes, environmental groups, and other organizations on energy markets, economic and environmental regulation and strategy, and climate-related energy policies. She has participated

APPENDIX A

as an expert in civil litigation cases, regulatory proceedings before state and federal agencies, and business consulting engagements. Previously, Dr. Tierney served as the assistant secretary for policy at DOE, and was the secretary for environmental affairs in Massachusetts, commissioner at the Massachusetts Department of Public Utilities, and executive director of the Massachusetts Energy Facilities Siting Council. She co-authored the energy chapter of the *National Climate Assessment*, and serves on the boards of ClimateWorks Foundation, Barr Foundation, Energy Foundation, Resources for the Future, and World Resources Institute. Dr. Tierney taught at the Department of Urban Studies and Planning at MIT and at the University of California, Irvine, and has lectured at Harvard University, University of Chicago, Yale University, New York University, Tufts University, Northwestern University, and University of Michigan. She earned her Ph.D. and M.A. in regional planning at Cornell University and her B.A. at Scripps College.

JENNIFER WILCOX is the Presidential Distinguished Professor of Chemical Engineering and Energy Policy at the University of Pennsylvania (UPenn) and leads the World Resources Institute (WRI) Carbon Removal Plan as a senior fellow. Dr. Wilcox joined UPenn and WRI following appointment as the James H. Manning Chaired Professor of Chemical Engineering at Worcester Polytechnic Institute. Having grown up in rural Maine, Dr. Wilcox has a profound respect and appreciation of nature, which permeates her work as she focuses on minimizing negative impacts of humankind on our natural environment. Dr. Wilcox's research takes aim at the nexus of energy and the environment, developing both mitigation and adaptation strategies to minimize negative climate impacts associated with society's dependence on fossil fuels. This work carefully examines the role of carbon management and opportunities therein that could assist in preventing 2°C warming by 2100. Carbon management includes a mix of technologies spanning the direct removal of carbon dioxide from the atmosphere to its capture from industrial, utility-scale exhaust streams, followed by utilization or reliable storage of carbon dioxide on a time scale and magnitude that will have a positive impact on our current climate change crisis. Funding for Dr. Wilcox's research is primarily sourced through NSF, DOE, and the private sector. She has served on a number of committees including at the National Academies and the American Physical Society to assess carbon capture methods and impacts on climate. Dr. Wilcox is the author of the first textbook on carbon capture, published in March 2012.

APPENDIX B

Disclosure of Unavoidable Conflicts of Interest

The conflict-of-interest policy of the National Academies of Sciences, Engineering, and Medicine (https://www.nationalacademies.org/about/institutional-policies-and-procedures/conflict-of-interest-policies-and-procedures) prohibits the appointment of an individual to a committee like the one that authored this Consensus Study Report if the individual has a conflict of interest that is relevant to the task to be performed. An exception to this prohibition is permitted only if the National Academies determine that the conflict is unavoidable and the conflict is promptly and publicly disclosed.

When the committee that authored this report was established, a determination of whether there was a conflict of interest was made for each committee member given the individual's circumstances and the task being undertaken by the committee. A determination that an individual has a conflict of interest is not an assessment of that individual's actual behavior or character or ability to act objectively despite the conflicting interest.

Dr. Ed Rightor was determined to have a conflict of interest in relation to his service on the Committee on Accelerating Decarbonization in the United States because he owns shares in Dow Chemical Company and DuPont.

Dr. Susan F. Tierney was determined to have a conflict of interest in relation to her service on the Committee on Accelerating Decarbonization in the United States because she is currently employed by a consulting company (Analysis Group) that provides analyses of energy markets, clean energy regulatory policy, and resource planning and procurement for a broad range of clients (including grid operators, utility and other energy companies, governments, nongovernmental organizations, and energy consumers) in the electric and natural gas industries.

In each case, the National Academies determined that the experience and expertise of the individual was needed for the committee to accomplish the task for which it was established. The National Academies could not find another available individual with the equivalent experience and expertise who did not have a conflict of interest. Therefore, the National Academies concluded that the above conflicts were unavoidable and publicly disclosed them on its website (www.nationalacademies.org).